CONTAMINATION
OF
ELECTRONIC
ASSEMBLIES

The Electronic Packaging Series

Series Editor: Michael G. Pecht, University of Maryland

Advanced Routing of Electronic Modules
Michael Pecht and Yeun Tsun Wong

Contamination of Electronic Assemblies
Michael Pecht, Elissa M. Bumiller, David A. Douthit, and Joan Pecht

Electronic Packaging Materials and Their Properties
Michael Pecht, Rakesh Agarwal, Patrick McCluskey, Terrance Dishongh, Sirus Javadpour, and Rahul Mahajan

Guidebook for Managing Silicon Chip Reliability
Michael Pecht, Riko Radojcic, and Gopal Rao

High Temperature Electronics
Patrick McCluskey, Thomas Podlesak, and Richard Grzybowski

Influence of Temperature on Microelectronics and System Reliability
Pradeep Lall, Michael Pecht, and Edward Hakim

Long-Term Non-Operating Reliability of Electronic Products
Michael Pecht and Judy Pecht

CONTAMINATION OF ELECTRONIC ASSEMBLIES

Michael Pecht, Elissa M. Bumiller,
David A. Douthit and Joan Pecht

CRC Press
Taylor & Francis Group
Boca Raton London New York

CRC Press is an imprint of the
Taylor & Francis Group, an **informa** business

CRC Press
Taylor & Francis Group
6000 Broken Sound Parkway NW, Suite 300
Boca Raton, FL 33487-2742

First issued in paperback 2019

© 2003 by Taylor & Francis Group, LLC
CRC Press is an imprint of Taylor & Francis Group, an Informa business

No claim to original U.S. Government works

ISBN-13: 978-0-8493-1483-4 (hbk)
ISBN-13: 978-0-367-39566-7 (pbk)
Library of Congress Card Number 2002031240

Library of Congress Cataloging-in-Publication Data

Contamination of electronic assemblies / Michael Pecht ... [et al.].
    p.   cm. (Electronic packaging series)
    Includes bibliographical references and index.
    ISBN 0-8493-1483-6 (alk paper)
    1. Electronic packaging--defects--Prevention. 2. Contamination (Technology) ..
    Factory sanitation. I. Pecht, Michael. II. Series.
TK7870.15 .C66 2002
621.381′046—dc21   2002031240
                                                                        CIP

Visit the Taylor & Francis Web site at
http://www.taylorandfrancis.com

and the CRC Press Web site at
http://www.crcpress.com

PREFACE

New components, designs, materials, and assembly processes have caused contamination problems to become a major factor in determining the manufacturability, quality, and reliability of electronic assemblies. Understanding the mechanics and chemistry of contamination has thus become necessary for increasing quality and reliability, and reducing the costs of electronic assemblies.

What this book is about

The purpose of this book is to address contamination issues in electronic products, with enough information to direct someone into the correct area with the ability to ask questions and understand answers for viable identification, mitigation, and management. This book is intended to be a guide to solve problems. It takes a step-by-step approach to identify the contaminants encountered at each level of processing and their detrimental effects on electronic products. The book discusses contamination of electronic assemblies from the manufacture of the glass fibers used in the laminates to the complete assembly of the finished product. Defects that are encountered during the manufacturing process are targeted, and the contaminants that cause defects and how these defects can be reduced by detection and control methods are discussed. Tables, figures, and fishbone diagrams serve as a quick reference to help engineers familiarize themselves with the origination, detection, measurement, control, and the prevention of contamination on electronic assemblies.

Who this book is for

Engineers employed in companies ranging from the laminate manufacturers to the end users can find a multitude of uses for the information and references contained in these pages. Among those most likely to find this information useful are:

- Design, process, reliability, system, and field engineers

- Quality assurance, students, managers, sales representatives
- Technicians, production workers, safety representatives

How this book is organized

This book is divided into three different parts: Laminate Manufacturing, Substrate Fabrication, and Printed Wiring Board Assembly. These three parts are then broken into subsequent chapters that identify the contaminant sources that occur during the process discussed, how to measure and control these contaminants in the process, and show how the contaminants can affect the quality and reliability of the final product. Then the cleaning methods that should be followed to limit or prevent contamination are provided.

Acknowledgements

The authors would like to thank the following people for their assistance in the preparation of this book. Without their help this work could not have been completed.

Thank you to:
Anoop Rawat for his tremendous assistance with editing and writing some of the conformal coatings material.
Mark Henry for the use of his Master's thesis work.
Keith Rogers for the use of the pictures and material on hollow fibers and conductive filament formation.
Craig Hillman for his guidance and assistance on the control methods chapters and for assisting in the editing of this book.
Ania Picard and Charles Lobo for their dedicated assistance in the formatting and production of this book.
James McLeish for his support and suggestions.

INTRODUCTION

Some definitions are presented here to introduce the concept of contaminates on electronic assemblies.

Contaminate – to render impure, undesirable, harmful, or unusable by contact with something unclean.

Contaminant – any matter, not designed into an electronic product, that is introduced during its manufacture or its use in the environment and is capable of unacceptably degrading product performance.

Contamination – occurs when any contaminant is present over the surface of, adsorbed on the surface of, or absorbed into the surface of electronic components, assemblies, or systems.

Not all foreign matter is harmful to electronic product performance. The matter may be a contaminant at one stage of manufacture, yet a desirable product at another stage. For example, during the print and etch process for manufacturing a bare copper printed circuit, the etch resist, if improperly stripped, becomes a contaminant in the soldering process– that is, the etch resist, an essential part of the manufacture of a printed circuit, becomes a contaminant during the soldering of the same printed circuit.

Although materials from a variety of sources can be present in a residue, they generally fall into three categories [43]:

Category 1: Particulates – resin and fiberglass debris, metal and plastic chips, dust, handling soils, lint, insulation, hair, skin

Category 2: Polar, ionic, inorganic – flux activators and residues, handling salts, plating and etching salts, neutralizers, fusing fluid residues

Category 3: Nonpolar, nonionic, organic – solid rosin, handling oils and greases, waxes, soldering oils, hand creams, polyglycol or polyol (polyhydric alcohol) degradation products, surfactants

Contaminants can be classified into seven classes [8]. These classes constitute a mixture of the three common categories of contaminants, so it is very difficult to assign a category for each class, and vice versa. Table 1 shows the different classes, examples, and their respective description. All or some of these contaminants can be present simultaneously, which can further complicate cleaning, because each type of residue may require a different cleaning material and process for effective removal.

Many substances belong to more than one class, either because they contain different contaminants, or because they consist of substances that can be removed by more than one method. Fingerprints, for example, contain both greases and salts, and to eliminate them may require a solvent and water– they thus combine classes A and C. On the other hand, fingerprints may be eliminated by a strong aqueous detergent solution alone, in which case they could be considered as class D. Therefore, it is sometimes simpler to describe contaminants as belonging to category 1, 2, or 3.

Table 1 Classification of contamination by solubility and mode of removal [8]

Class	Examples	Description
A	Oils and greases	Most soluble in hydrophobic solvents such as chlorinated, fluorinated, and certain aromatic and aliphatic solvents (trichloroethylene, F-113, toluene, and benzene).
B	Rosin	Most soluble in hydrophilic solvents such as alcohol, ketones, acetone, methyl-ethyl ketone.
C	Salts and sugars	Most soluble in water.
D	Metal oxides, oils and greases, other soils	Most soluble in special aqueous media, such as hydrochloric acid mixed with water or a detergent mixed with water.
E	Dust and larger particles	Insoluble, but removable by mechanical means such as blowing, aspiration, displacement in a liquid, or brushing.
F	Metal between two conductors	Insoluble, difficult to remove by mechanical means such as abrasion or scraping with a knife.
G	Epoxy smears, anything under the surface	Cannot be removed.

The classifications are very broad and there are some twenty or thirty different sub-classifications. For example, one of the most important distinctions is between ionic and nonionic class C contaminants. If a small quantity of salt is added to pure water at 20°C, the resistivity of the water will drop very rapidly, illustrating that the salt is ionic. On the other hand, if a small quantity of sugar is added to pure water at 20°C, the resistivity of the water will not change, since sugar is nonionic. Even though both products dissolve similarly in water, they produce different results [8].

Different kinds of ionic products produce differently characterized solutions. For example, some ionic products, such as sodium chloride, are very readily soluble in water and produce very low resistivities with small quantities [143]. Other kinds of ionic products, such as lead carbonate, may dissolve much less readily in water. Above all, the same weight of different products will produce different resistivities. Moreover, the voltage/current characteristic is not linear, particularly at low voltages [144], because it requires a certain threshold voltage to tear the positive and negative ions apart; at the same time, their inherent charges tend to keep them together.

A summary of the sources of contamination is given in Figure 1. This diagram can be used in two directions. From outside in, the resultant contaminants can be traced based on the process step. From the inside out, the diagram can trace the possible source(s) of contamination. In the processing illustrated in this diagram and more detailed ones that follow, some contaminants undergo soldering operations at temperatures over 250°C. The lowest level in each diagram indicates by-products, but not all the original contaminants are modified by heat. It is normal to presume that after soldering, there will be contaminants present from both the lowest and higher levels.

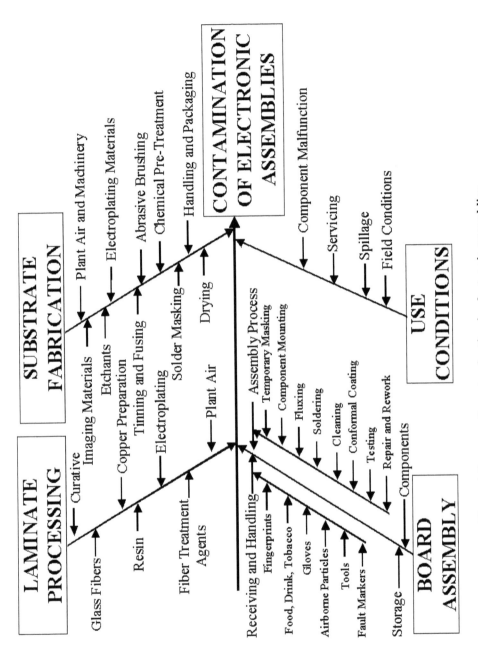

Figure 1 Sources of contamination in electronic assemblies

TABLE OF CONTENTS

PART 1

LAMINATE MANUFACTURING

CHAPTER 1   CONTAMINANT SOURCES IN LAMINATES

A substrate is a platform on which electronic components are mounted. Also known as a printed wiring board (PWB) or a printed circuit board (PCB), a substrate provides mechanical support for components and conductors, insulates the routing signals, and provides a conduction path for heat dissipated by components. The base material of the substrate is the laminate. Four common categories of organic laminates☐ rigid, flex, rigid-flex, and molded☐ can be constructed into either a single-sided, double-sided, or a multi-layered substrate.

The laminate is a composite covered by metallization. Composites are made up of reinforcements and a resin matrix. The reinforcement is the "backbone" of the substrate. It provides multiple paths for energy dissipation that help the laminate to resist damage and, in electronics, is usually glass fiber. The resin matrix provides protection for the reinforcements, provides a continuous surface, and allows for load transfer between the fibers via shear. The matrix is usually an organic binder. The metallization is the metal used for routing signals. Some common materials used in the electronics industry for each of these are listed below.

∞ *Reinforcements* – E-glass, C-glass, A-glass, S-glass, D-glass, Quartz, and aramids

∞ *Resins* – FR-4 epoxy, multifunctional FR-4 epoxy, bismaleimide and triazine (BT), cyanate ester (CE), polyimide, and polytetrafluoroethylene (PTFE)

∞ *Metallization – copper, aluminum, silver, and gold*

Since there are numerous categories, types, and materials, many variations of laminates exist. When the term laminate is used in this document, it refers to a rigid plastic laminate. A typical example could be FR-4 epoxy matrix, E-glass reinforcement with copper metallization.

The manufacture of the composite of a laminate can be broken down into three simple steps (Figure 1.1). The first step is the manufacture of glass fibers through a fiber-drawing process. The second step is producing either a woven or nonwoven fabric. The third step is applying an A-stage liquid resin to the fabric sheets and partially curing them via heat and pressure to form a "prepreg."

Numerous layers of prepreg are stacked and pressed together to form a laminate. There are two methods of lamination available to manufacturers: hydraulic lamination and vacuum lamination. With hydraulic pressing, high heat and high pressure are used to cure materials so that there is a homogenous mix and to get rid of the air trapped between prepeg layers. However, with the high-pressure system, there is a large amount of stress on the board. Therefore, many manufacturers have to post-bake the board in order to relieve the built-in stress. Also, with hydraulic pressing, "footballing" occurs. The middle of the board gets thicker than the edges. To prevent this anomaly from occurring, manufacturers use the more expensive method of vacuum pressing. Instead of getting rid of air through high pressure, a vacuum is created and a positive amount of pressure (around 90-110 psi compared to the 350 psi used in the hydraulic pressing) is placed on the system to keep everything together. The result is a flat board with the same thickness with which it started [83].

Some laminates include a layer of copper cladding on one or both sides. The copper cladding eventually becomes two inner layers, with the laminate acting as dielectric spacing between the layers [1]. The following sections briefly describe some of the contaminants found in the materials and equipment used in laminate fabrication.

Figure 1.1 Typical process flow for laminate fabrication

1.1 Glass Fibers

Chemically, glass is composed mainly of silicon dioxide (SiO_2). Pure SiO_2 requires a very high temperature (1700℃) to be melted. Other chemical components are added to change its properties so it can be melted, homogenized, evacuated of air bubbles, and drawn into fibers.

For applications requiring excellent electrical properties, an E-glass composition is used. This glass is chemically a calcium alumino-borosilicate material with a low alkali content. E-glass fiber is the material most commonly used as reinforcement for the FR-4 substrates in the electronic industry because it has a high thermal and chemical resistance, low moisture absorption, and a high dielectric and tensile strength at a relatively low cost [1]. The typical chemical compositions of E-glass are given in Table 1.1.

Table 1.1 Typical chemical compositions of the E-glass fibers [3]. With permission from the Institute for Interconnecting and Packaging Electronic Circuits.

Constituents	Weight %
Silicon dioxide (SiO_2)	52-56
Aluminum oxide (Al_2O_3)	12-16
Calcium oxide (CaO)	16-25
Magnesium oxide (MgO)	0-5
Boron trioxide (B_2O_3)	5-10
Ferric oxide (Fe_2O_3)	0.05-0.4
Sodium oxide (Na_2O)	0-2
Potassium oxide (K_2O)	0-2
Fluorine (F_2)	0-1
Titanium oxide (TiO_2)	0-0.8

1.2   Fiber Treatment Agents

When the glass fibers are prepared for fabric manufacture, they are treated with a silane coating to produce a good structural bond to the epoxy resin. The glass fiber itself generally causes no contamination, but the coatings can create contamination if the glass fiber is exposed for any reason, such as poor pressing, excessive use of abrasives, or chemical attack of the epoxy [12].

The treatment agents used for this coating usually contain a silicone or chromium group (silane, chromic chloride) that reacts with the fiber surface, and an unsaturated organic group (amine, allyl resorcinol, vinyl, methacrylate) that reacts with the functional radicals of epoxides, polyesters, and other resins [5], as shown in Table 1.2.

Table 1.2 Common agents for glass fiber treatments [5]

Resins	Common Treatment Agents
Phenolic	Amino silane
	Allyl resorcinoxy silane
Melamine	Amino silane
Polyester	Vinyl silane
	Methacrylate chromic chloride
Epoxide	Amino silane
	Methacrylate chromic chloride
Silicone	Vinyl silane

Some of the treating agents, in particular methacrylate chromic chloride and some of the silanes that are soluble in aqueous media, are capable of producing vesication or mealing, which is the blistering of the laminate or conformal coating [4]. Vesication begins when the outer layer of resin is removed via prolonged exposure to harsh solvents, exposing the glass fiber. The treatment agents react with moisture and degrade the surface resistance of the laminate. When a protective coating is applied to the board, the vaporization of the absorbed moisture causes blistering [14].

1.3   Epoxy Resin

The resin is the binder that holds the laminate together. During the fabrication of the laminate, the resin goes through three stages. A-stage resin is a liquid resin that is mixed and catalyzed, but not cured. B-stage resin is heated and partially cured during the treating process; this resin is dry to the touch. C-stage resin is fully cured during the pressing process. This epoxy resin is also a potential source of contamination [13].

The resin mixture must be free from foreign material and relatively low in moisture. Any entrained material in the resin that can explosively decompose or volatilize during high-temperature soldering operations or high-temperature lamination conditions must be avoided. Entrained moisture in the mixture can also negatively impact the electrical performance (dielectric constant and dissipation factor) of the resultant laminate.

In the thermosetting resin systems used in FR-4 boards, the resin is polymerized using a curative (a hardening agent). The weight ratio of resin to curative must be specific and exact for each system to yield as complete a reaction as possible and leave minimal excess. Any excess of unreacted resin will be embedded in the cured resin, resulting in increased moisture absorption, reduced solvent resistance, a reduced glass transition temperature, or pronounced degradation of electrical properties with increased temperatures.

If there is an excess of curative, such as dicyandiamide, the dielectric properties of the laminate will degrade due to the polar properties of the typical amines used as hardening agents [4]. Amines are a class of organic compounds of nitrogen derived from ammonia (NH_3), in which one or more of the hydrogen atoms are replaced by organic radicals, such as CH_3 or C_6H_5. All amines are basic in nature, and will usually combine readily with hydrochloric or other strong acids to form salts [6]. If there is too much or too little hardener, then the resin will leave contamination that is nearly impossible to remove and partially ionic. This dangerous condition is fortunately fairly rare with established laminate manufacturers, but the problem is also aggravated when additives are used to alter the basic characteristics of the epoxy resin. One example is the bromide additive used to render the epoxy flame retardant in FR-4 boards (FR stands for flame retardant). If this additive is not correctly cross-linked into the epoxy resin, the resultant contamination is strongly ionic and the electrical characteristics of the laminate are seriously compromised [4]. Other additives that could have a similar effect are those used for increasing either the glass transition temperature (T_g) of the resin and its dimensional stability or temperature resistance characteristics.

One method of determining if the resin system has been properly cured is to mechanically peel the copper off a panel and then perform an ionic contamination test. If the panel shows even a small amount of ionic contamination, it should be considered suspect for applications requiring good insulation resistance, although it may be perfectly acceptable for low-impedance applications.

1.4 Copper Cladding

The copper cladding can be a source of contamination during laminate manufacture, frequently causing poor solderability. The main cause of copper contamination is epoxy bleed through the pores between the individual crystals of copper. Similarly, the vapors formed during the laminating process can carry organic contaminants through the copper [11]. This problem is aggravated by brushing, which tends to smear any resin bleed-through over the whole surface of the copper.

The copper cladding process is another potential source of contamination, depending on whether rolled or electrolytic copper is used. Rolled foil, produced by conventional metal rolling techniques, has less chance of being contaminated with inclusions and foreign metal. Electroplating the copper from an acid copper sulfate or alkaline copper cyanide bath onto stainless steel or leaded drums produces the electrolytic copper. After the electroplated copper foil deposit is stripped from the drum, the side exposed to the metal cathode can be contaminated with the cathode material, usually lead. If this metal impurity protrudes to the outer surface of the foil, the plated deposits applied during board fabrication may not properly adhere to the leaded spots. Other possible results are pinholes formed during circuitry etching and resin bleed through the lamination process itself [4].

In order for the copper foil to bond well to the laminate, the foil must be prepared by chemical, electrolytical, or physical means. The surface of the foil can be oxidized by chemical, electrolytical, or thermal treatments [9]. Other methods include priming with organometallic compounds, such as titanates, and flash coating with other metals. The oxides, metal flash coats, and primers must be removable during the subsequent chemical copper etching process, or they will become contaminates for the finished circuitry.

1.5   Volatile Matter

Contaminants from laminate materials or processes can result in matter that is easily vaporized, called volatile matter. Air, water, or solvents used in the preparation of the resins can be trapped between layers. If air or solvents become trapped, high temperatures or voltages can cause internal delamination from increased vapor pressure. If moisture gets trapped inside the laminate, then "measling," a separation of the resin from the fiber, results with serious degradation of insulation resistance under high humidity conditions. Good inspection procedures, with adequate acceptance testing, can detect a laminate that is excessively contaminated with volatile matter or is highly porous early in the process step when rejection cost is minimal.

1.6   Summary of Contaminants

Detrimental or not, foreign matter can originate from various sources in the manufacturing of laminates. Whenever a new process step, a new material, or a new piece of equipment is introduced to the process, contaminants will follow. The contaminants and their consequences during the previous version of laminate processing are summarized in Table 1.3.

Table 1.3 Summary of potential contaminants from laminate processing

Contaminant Sources	Contaminants	Consequences
Glass fibers	Boron, calcium, aluminum, silicon, magnesium, sodium, potassium, titanium, iron, fluorine	Vesication Moisture absorption Degradation of insulation resistance
Fiber treatment agents	Amino silane, allyl resorcinoxy silane, vinyl silane, methacrylate chromic chloride	Vesication Moisture resistance degradation Degradation of insulation resistance
Resin	Phenolic, melamine, polyester, epoxide, silicone, bromide, resin bleed-through	Increased moisture absorption Volatilization Reduced solvent resistance Reduced glass transition temperature Degradation of electrical properties Poor adhesion
Curative	Amines, dicyandiamide	Degradation of electrical properties Salt formation Poor adhesion
Electroplating	Lead, copper sulfate bath, stainless steel, foreign metals	Poor adhesion Pinholes Resin bleed
Copper preparation	Oxides, metal flash coats, primers	Poor adhesion Pinholes
Plant air	Dust, dirt, dander, foreign material	Poor bonding Poor adhesion Insulation resistance degradation

CHAPTER 2   MEASURING CONTAMINANTS IN LAMINATES

According to IPC-EG-140, Specification for Finished Fabric Woven from 'E' Glass for Printed Boards, the only contaminants considered are foreign inclusions from airborne materials or direct contact during processing. These defects (defined as "*a substandard area in the fabric*") are "*clearly noticeable*" and are considered a major defect (defined as "*a defect that is likely to result in failure, or to reduce materially the usability of the unit of product for its intended purpose*"). "*The acceptable size and frequency of foreign inclusions shall be agreed upon between buyer and seller [3].*" This means that there are no industry standards in place for any other contaminants, such as ionic or nonionic, except for particulate matter. In fact, there are no established cleaning methods to eliminate foreign inclusions. The material is accepted and used, or it is not accepted and discarded.

Only a few manufacturers perform some type of testing during their fabric manufacturing process, and they test predominately for moisture content. The moisture is measured by the fabric weaver with a loss-on-ignition (LOI) test, which measures weight loss after high-temperature baking (usually above 315°C). The LOI test, typically performed before applying any silane treatment agents, has a target value of less than 0.1 percent weight loss. The treatment agents are applied ("finished") at a very low level (0.02 percent), according to IPC-EG-140. After the fabric is finished, the only contaminants the manufacturer considers are environmentally acquired, via air or direct contact. These contaminants are not measured at this point due to their randomness, and are controlled by clean rooms.

Some manufacturers assess the moisture content of the resin mixture using the Karl-Fisher titration test. Moisture content must be maintained below approximately 0.25 percent. Foreign material, usually acquired during mixing operations or entrained in the raw material, is typically minimized by filtration processes, but is not quantified.

The laminating process is sensitive to incompletely removed solvents as contaminants and to any environmental contaminants, but the acceptable levels of residual solvent are not quantified. However, some manufacturers are conscious of the occurrence and impact of these contaminants and take appropriate steps to minimize them. One step is to implement a continuous lamination process. In continuous lamination, the thermosetting resin mixture is free of solvents, the reinforcement is impregnated with the resin mixture, and the layers of reinforcement and copper foil are laminated and cured from liquid to fully cured laminate in a continuous process. The continuous lamination process is sensitive only to environmentally acquired contaminants.

Manufacturers strive to minimize environmental contaminants through clean rooms, but unfortunately, contaminants can be introduced and no tests are implemented to detect them. Since there are no formal standards for measuring contaminants, glass fabric and laminate manufacturers can get away with producing and selling defective products. Unfortunately, these defects may not be observed until after the boards are assembled, or even later, when failure analysis is performed.

Some measurements made during the manufacturing process can indirectly identify possible contaminants. For example, it is a best practice to measure laminates for flame retardant content and resin content, both of which if present in excess amounts can become contaminants later in the process. Flame retardants can be particularly harmful because they contain bromine. Bromine, in excess, is free to bond with moisture and cause dendritic growth or conductive filament formation problems.

CHAPTER 3   QUALITY AND RELIABILITY OF LAMINATES

The following sections illustrate some of the defects that can arise from contamination at the laminate level. They are divided into two general defect groups: externally observable and internally observable. Externally observable defects are defined in the IPC-A-600 standard titled <u>Acceptability of Printed Boards</u> as "those features or imperfections, which can be seen and evaluated on or from the exterior surface of the board. In some cases, such as voids or blisters, the actual condition is an internal phenomenon and is detectable from the exterior." The standard defines internally observable defects as "those features that required micro-sectioning of the specimen or other forms of conditioning for detection and evaluation. In some cases, these features may be visible from the exterior and require micro-sectioning in order to assess acceptability requirements [72]." Both groups are discussed in this section with figures illustrating identification of the defect externally and internally.

3.1 Weave Exposure and Weave Texture

Weave exposure and weave texture are two conditions that arise on the laminate surface. Weave exposure (Figure 3.1) occurs when unbroken fibers of woven glass cloth are not completely covered by the resin. This condition could decrease the dielectric properties of the laminate and cause resin reversion, as well as initiate a path for conductive filament formation.

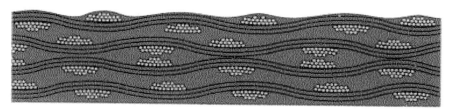

Figure 3.1 Illustration of weave exposure on a glass laminate. Notice that some glass fibers are completely exposed [72]. With permission from the Institute for Interconnecting and Packaging Electronic Circuits.

Weave texture (Figure 3.2) is similar to weave exposure, but a thin layer of resin covers the unbroken glass fibers.

Figure 3.2 Illustration of weave texture on a glass laminate. The resin completely covers the glass fibers, but is not sufficiently thick to provide a smooth flat surface [72]. With permission from the Institute for Interconnecting and Packaging Electronic Circuits.

3.2 Pits and Microvoids

Pits, also called microvoids, are small air pockets under the surface of a laminate (Figure 3.3). Pits can reduce the dielectric properties of the laminate, as well as allow for additional moisture to be absorbed into the surface. When the laminate is made into a substrate, conductor traces may be created within the voids; moisture will provide a conductive path leading to an electrical short. The moisture will also tend to vaporize during soldering and produce more serious imperfections leading to greater reliability problems.

3.3 Measling and Crazing

Measling and crazing are conditions which produce a separation at the fiber/resin interface (see Figure 3.4). Measling is a separation of the glass fibers at the intersection of the weaves, while crazing is a separation of the resin from the fibers at the same location. These defects manifest themselves as white spots below the surface of the base materials. When they become connected, "delamination" occurs. Since these conditions are strictly subsurface phenomena and occur as a separation of fiber bundles at fiber bundle intersections, their positions relative to surface conductors have no apparent significance. However, the separation of the fiber/resin interface leaves a path for conductive filament formation to initiate and can change the dielectric properties of the electrical system.

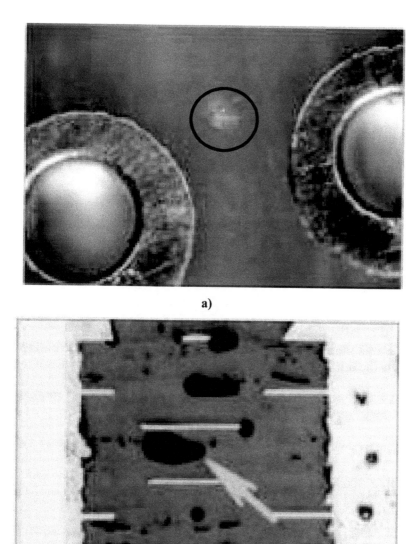

a)

b)

Figure 3.3 a) Photograph of a cluster of pits under the surface of a laminate,
b) Cross-section of voids between layers [72]. With permission from the
Institute for Interconnecting and Packaging Electronic Circuits.

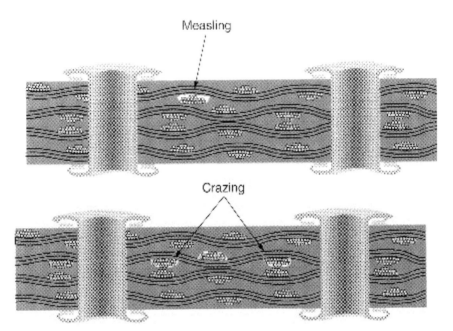

Figure 3.4 Illustration showing measling and crazing [72]. With permission from the Institute for Interconnecting and Packaging Electronic Circuits.

3.4 Blistering and Delamination

Blistering (Figure 3.5) is a separation between any of the layers of the base material or between the base material and metal cladding. Blistering can lead to delamination of adjacent layers, which will degrade the thermal and electrical properties of the dielectric and can cause electrical shorts.

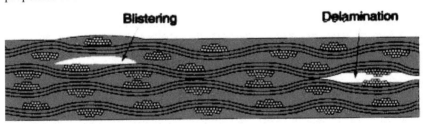

Figure 3.5 Illustration of blistering and delamination [72]. With permission from the Institute for Interconnecting and Packaging Electronic Circuits.

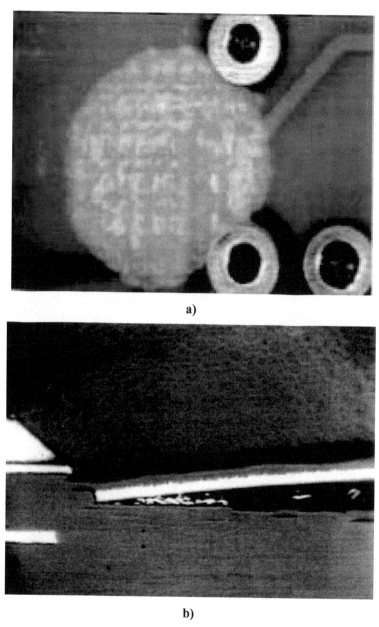

a)

b)

Figure 3.6 a) Top view of a blister between two plated through holes, b) Cross-sectional view of delamination at a copper trace interface [72]. With permission from the Institute for Interconnecting and Packaging Electronic Circuits

3.5 Foreign Inclusions

Foreign inclusions (Figure 3.6) are metallic or on-metallic materials that are trapped in the laminate. They may be detected in the raw laminate, B-stage, or processed multiplayer boards, and may be conductive or non-conductive.

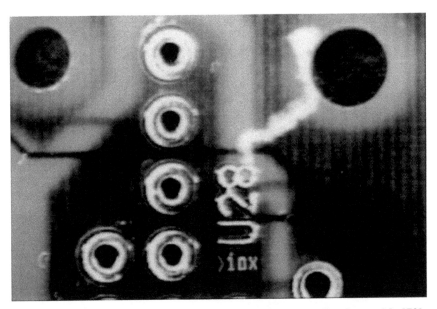

Figure 3.7 Top view of a foreign inclusion that is externally observable [72]. With permission from CALCE Electronic Products and Systems Center.

3.6 Hollow Fibers

A hollow fiber (see Figure 3.8 and Figure 3.9) if exposed on both sides can absorb contamination via capillary action from chemicals used in the process. Chemicals will reside inside the open path, awaiting convenient condition to cause conductive filament formation (CFF) between conductors (see Figure 3.10). Residues inside the hollow fibers are extremely difficult to remove, as the fiber size is on the micro scale.

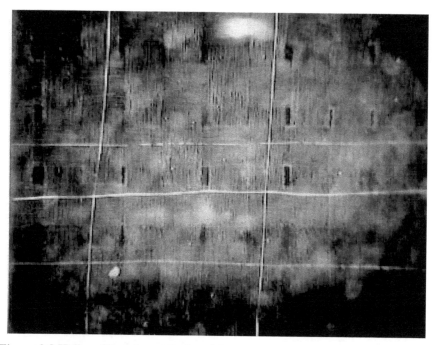

Figure 3.8 Hollow fibers in vertical and horizontal directions. With permission from CALCE Electronic Products and Systems Center.

Figure 3.9 Hollow fiber cross-section. With permission from CALCE Electronic Products and Systems Center.

3.7 Conductive Filament Formation

Conductive filament formation is an electrochemical process that involves the transportation (usually ionically) of a metal through or across a non-metallic medium under the influence of an applied electric field [62-64]. CFF can result in either leakage currents that reduce performance or catastrophic shorts that cause complete failure. The biased conductors act as electrodes providing a driving potential, while ingressed moisture between the organic resin and the fiber reinforcement will serve as an electrolyte (see Figure 3.10). As metallic ions migrate and form a bridge between two biased conductors, the loss of insulation resistance results in a current surge. The current surge will eventually cause a short and a large increase in localized temperature, which can manifest itself as a burnt or charred area between the two conductors (Figure 3.11).

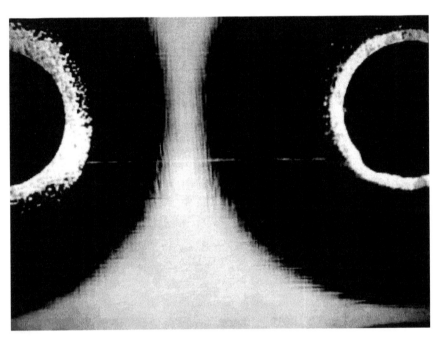

Figure 3.10 Two plated through holes shorted by a copper filled hollow fiber. With permission from CALCE Electronic Products and Systems Center.

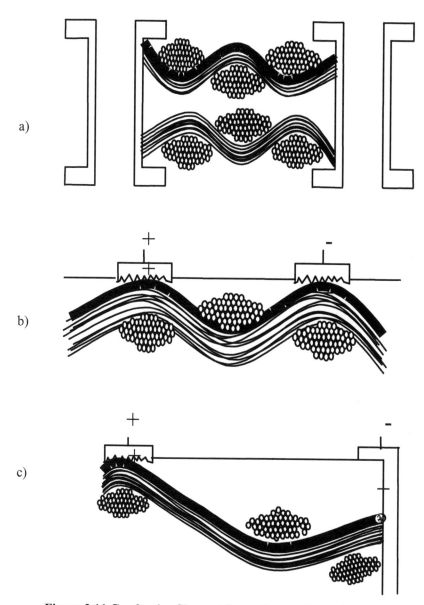

Figure 3.11 Conductive filament formation configurations:
a) between two plated through holes, b) between two surface traces,
and c) between a surface trace and a plated through hole. With permission
from CALCE Electronic Products and Systems Center.

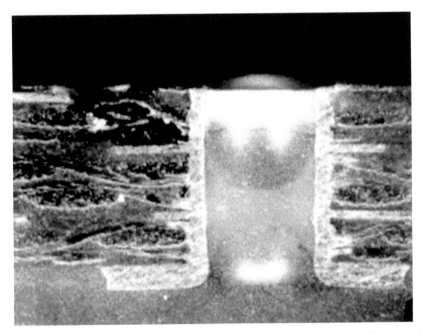

Figure 3.12 A plated through hole and power plane shorted by CFF charred the immediate area. With permission from CALCE Electronic Products and Systems Center.

The prime factors influencing CFF are board features (resin, materials, conformal coatings, and conductor architecture) and operating conditions (voltage, temperature, and relative humidity). Path formation, necessary for CFF, will occur due to interfacial delamination of the interface between the individual fibers and the organic resin matrix (Figure 3.12). Poor drilling and thermal cycling often precipitate this degradation. Previous studies have shown the quantitative effect of these factors on the occurrence of electromigration [65-67]. Models have been presented that predict the operational lifetime, assuming an eventual failure mechanism of CFF [67, 68].

More recent experiments have shown that CFF can also occur in the presence of hollow fibers [69, 70]. The environmental considerations are the same, but in this case the path formation occurs within the fiber itself, instead of along the fiber/matrix interface. The scenario of electromigration within the fiber and its effect on time to failure has also been modeled [71].

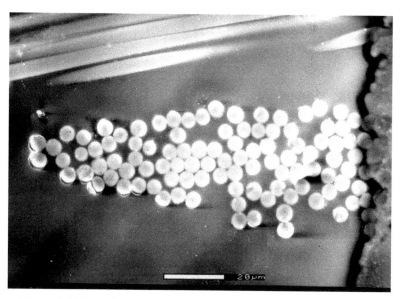

Figure 3.13 Interfacial degradation (debonding) at the fiber/epoxy resin interface

3.8 Resin Reversion

It is well known that the hardening of epoxy resins is due to their polymerization. What is perhaps less known is that all polymeric substances can, under certain conditions, revert into their monomers or at least into lower molecular weight compounds [4]. This is most common with elastomeric compounds (silicone rubbers), and less so with the harder polymeric compounds, such as epoxies and similar plastic materials. Reversion increases in the presence of humidity and at higher temperatures, although there is an optimum temperature and humidity condition, probably around 95 percent relative humidity at 95°C [4]. If reversion takes place, even partially, both the mechanical and electrical characteristics of the resin will be affected. In the case of epoxy resin, the type of hardener used has a significant effect. Amine reacted epoxy systems seem to be much more resistant than systems that have been catalyzed with polycarboxylic acids and hydrides, or a mixture of amine and acid hydrides. It is the responsibility of the laminate or resin manufacturer to prevent reversion. However, the probability of polymer reversion occurring in a printed circuit during normal conditions is very slight, the only exception to this generalization is when the circuit is encapsulated with an elastomeric substance, such as silicone rubber or a similar material.

CHAPTER 4   CONTROL METHODS IN LAMINATE
MANUFACTURING

Control of the laminate manufacturing process is essential to minimize contamination of product. At this stage in the printed circuit board assembly process, the only real protection method that is enforced is the use of clean rooms during the manufacture of the laminate. However, certain control methods should be in place to monitor the contaminants present in the manufacturing process and to ensure a quality product is being produced.

4.1   Manufacturing Process Assessment

The assessment of the manufacturing process through proper control items can help to identify if a process is releasing contaminants into the product. There are currently no industry-wide standards to grade or assess the manufacturing processes of laminate manufacturers. However, IPC-EG-140 Specification for Finished Fabric Woven from 'E' Glass for Printed Boards provides suggestions on glass fabrics. The following are suggestions to guide the control monitoring of the manufacturing processes of laminates.

4.1.1 Quality Assurance Through Certification

In order to monitor processes that may allow contamination into a laminate, quality assurance through certification or external audit should be considered. This encompasses three areas: state what you are going to do, do what you state, and document what you do.

4.1.2 Statistical Process Control

The measurement of critical parameters within the manufacturing process can identify when a process or group of processes are not performing acceptably. This may also indicate potential contamination sources.

4.1.3 Optimization

The optimization of several key processes within the manufacturing process may also help to prevent contamination. Some examples are optimization of the resin smear process, of the prepreg treating process, of the copper foil treatment process, and of the prepreg copper foil bonding process. This list is not meant to be exhaustive in nature, but attempts to provide common examples of how processes that are not optimized may expose the laminate to contaminants.

The resin smear process is the process when the resin is smeared onto the glass fabric in order to make the process. This process may be affected by changes in the resin, such as a resin with a higher viscosity than was previously used. It is important to optimize this process to the particular resin being used so that there will not be excess resin in the prepreg, which may prove to be a contaminant later in the printed wiring assembly process.

The prepreg and or copper foil may be chemically treated to make bonding together easier. The chemicals used in this process should be considered when optimizing this process along with the bond strength. Optimization would include the right amount of chemicals.

4.1.4 Cleaning

Cleaning methods vary from manufacturer to manufacturer. It should be noted that incomplete removal of cleaning solvents and/or residues leaves detrimental contaminants on the laminate, reducing the reliability of the finished assembly.

4.1.5 Best Practices

Best practices are usually not required for acceptable products, but are strongly recommended for quality product. With this in mind, a list of best practices to monitor and prevent contamination in laminates is given in section 4.4.

4.2   Control Monitoring Through Testing

Testing in laminate manufacturing is not common because testing is expensive and takes time. However, some tests recommended below should be performed to monitor the quality of the laminate.

4.2.1 Visual Inspection

Visual inspection of the laminate can find several problems such as foreign inclusions, blistering, voids, or delamination. While it may be impossible to 100 percent visually inspect every laminate, care should be taken so that the uniformity of the product is not in question.

4.2.2 Glass Transition Temperature Testing

A consistent glass transition temperature can be an important factor in the process control of a laminate. Temperatures that are out of the acceptable range could be the result of contamination in the glass fibers or the resin. Each lot of laminate should be examined for appropriate glass transition temperature.

4.2.3 Flame Retardant Content Measurement

Flame retardants are extremely helpful in the prevention of fire in electronics. Flame retardants usually contain halides, more specifically, bromides. In excessive amounts, these ionics can prove extremely harmful for the later product, causing electrochemical reactions such as dendritic growth, conductive filament formation, and corrosion.

4.2.4 Prepreg Resin Content Measurement

As with flame retardant, resins are generally a positive part of the laminate, but in excess amounts may cause contamination later in the product life. Too low of a resin content may also allow contamination to enter in the glass fabric. The resin content should be monitored and maintained within proper control limits.

4.2.5 Copper Foil Surface Inspection

Pits, dents, scratches, or other surface imperfections, such as discoloration, in the copper foil may be indicators of contamination in the copper foil manufacturing process. Each lot of copper foil should be inspected for these imperfections.

4.3 Root Cause Analysis

Root cause analysis procedures should be in place for when a noncomformity is found in the product. These procedures must be based on physics of failure (POF), so the failure mechanism is correctly identified and the problem corrected. There are whole books written on root cause failure analysis; therefore, the following section will cover only these topics as they relate to contamination.

4.3.1 Material Traceability

Materials should be marked in such a way as to know manufacture date, time, and lot number. Contamination is lot sensitive in many cases, making materials traceability an extremely important part of the root cause analysis.

4.3.2 Corrective and Preventive Actions

Once the nonconformity is found and traced to the root cause, a procedure for corrective actions should be in place so there are guidelines already in place to quickly correct the problem. These procedures can directly relate to guidelines for preventive actions, so once a nonconformity occurs and is corrected, it will not happen again.

4.3.3 Documentation

A tool available for corrective and preventive actions is to create a contamination database which records possible sources of contamination and the proper method for correction. This also allows for the experience of employees to be passed to those with less experience.

Process Change Notification. It is important to understand that any change in the laminate manufacturing process may introduce chemicals that become contaminants later in the printed wiring assembly. Therefore, it is paramount that process change notifications are available to all customers. This is true whether it is the copper foil supplier changing the electroplating bath or a change in the resin used in the prepreg, or any other process that adds or subtracts, chemicals, handling, etc. Process changes should be recorded with date, time, the process changed, and any other information that may be pertinent, such as reason for change.

4.4   Supply Chain Management

If a laminate is to be free from contamination, the supplies used to make the laminate should be free from contamination, as well as any storage, handling, and shipment materials. Assessment of the vendors from which the supplies are procured is recommended. Well-documented and careful storage, shipping, and handling is also recommended.

4.4.1 Vendor Assessment

The raw materials (glass fabric, resin, and copper foil) are usually supplied by outside vendors, who need to take matters into their own hands with regard to contamination of their product. Although some standards are in place for the contaminants discussed previously, more are needed. Each manufacturer is required to make sure its product is contamination free and shipped with care. The following tables provide questions that can be asked to assess the vendor's process and capability. The list is not meant to be exhaustive; rather it is meant to give a starting point on how to monitor contaminants from the beginning of the process.

Table 4.1 Assessment questions for glass fabric vendors

DEFECT/FAILURE MECHANISM	POLICY/PROCEDURE	YES	NO
Hollow Fibers	Does the supplier of glass fabric/plies screen for hollow fibers on every fabric purchase lot?		
Fiber/Resin Delamination	Is a qualification test performed on prepreg or laminate coupons to ensure no fiber/resin delamination will occur during assembly?		
	Is this qualification test performed for every prepreg or laminate purchase lot?		

Table 4.2 Assessement questions for epoxy/resin vendors

DEFECT/FAILURE MECHANISM	POLICY/PROCEDURE	YES	NO
Excessive resin smear caused by excessive melt viscosity	Does the resin supplier measure the melt viscosity on each batch of resin?		
	Does the board manufacturer re-optimize the desmearing process with every change in resin composition or resin supplier?		
Glass transition temperature (T_g) below specification	Does the resin supplier measure T_g on each batch of resin?		
Excessive volatile content	Does the resin supplier measure each batch of resin for volatile content?		
	Does the board manufacturer re-optimize the lamination process with every change in resin composition or resin supplier?		
Flammability resistance is low	Does the resin supplier measure each batch of resin for flame-retardants content?		

Table 4.3 Assessment questions for copper foil vendors

DEFECT/FAILURE MECHANISM	POLICY/PROCEDURE	YES	NO
Opens caused by surface imperfections	Does the copper foil supplier screen and document surface imperfections (pinholes, pits, scratches, wrinkles, inclusions) on each lot of copper foil?		
Delamination at the copper/prepreg interface due to improper surface treatment of copper foil	Does the copper foil supplier optimize the copper foil bonding treatment process?		
	Is the process re-optimized for every change in the copper surface treatment process?		
Mechanical properties below specifications	Does the copper foil supplier perform ductility and tensile strength testing on each lot of copper foil?		

Table 4.4 Concerns for prepreg vendors

DEFECT/FAILURE MECHANISM	POLICY/PROCEDURE	YES	NO
Insufficient or excessive resin content	Does the prepreg supplier optimize the treating process? Optimization will include understanding how resin flow and gel time affect resin content.		
	Does the prepreg supplier perform resin content testing on each lot of prepreg?		
Insufficient resin curing	Does the prepreg supplier measure the degree of cure for every lot?		
	Does the board manufacturer re-optimize the lamination with every change in prepreg supplier?		
Excessive moisture absorption	Are prepregs stored in controlled, dry (<40%) conditions or sealed in moisture proof packaging?		

4.4.1.1 Consumables Applied During Manufacturing

Any materials applied to the product, machinery that touches the product, or that can evaporate into the atmosphere should be documented and the contents of the material should be known.

4.4.2 Shipping, Handling, and Storage

Careful handling procedures are required during shipping of the laminates to the substrate manufacturer. Fingerprints, lint, dust, moisture, and other atmospheric contaminants adhere to the laminate. If no cleaning is incorporated before multi-layering, these contaminants can become trapped and are difficult to remove, if they can be removed at all. The handling and packaging of laminates should be given as much attention as the substrates and finished assemblies.

PART 2

SUBSTRATE FABRICATION

CHAPTER 5   CONTAMINANT SOURCES IN SUBSTRATES

Contamination in the substrate can originate in the fabrication process (Figure 5.1). Substrates are usually fabricated with one of several variations: single- or double-sided without plated holes, single- or double-sided with plated holes, and multi-layer with plated holes. These printed circuits (etched circuits) are produced via a sequence of operations that differ slightly for the different substrates.

Nearly every operation in the manufacture of a substrate can contribute to contamination. Good contamination control throughout the process is essential for making high-quality boards. This section presents the more common causes of contamination and what can be done to prevent them.

5.1   Plant Air and Machinery

Periodic monitoring of common industrial pollutants is advisable. These ions, such as SO_2, NO_2, hydrogen chlorides, ammonia, ozone, will vary widely depending on the plant location and the time of year. These ions are capable of causing corrosion and oxidation of metals quite rapidly if the humidity is over 40%. Moisture will begin to build on exposed metal surfaces up to 2 molecular layers at 40% RH, 2-4 molecular layers at 60% RH, and 5-10 layers at 80% RH [85]. Temporary and long-term storage, as well as the work area, needs to be monitored to prevent unnecessary rework or failures.

Nearly all of the mechanical operations employed in substrate manufacture can and do cause a certain amount of contamination. Almost inevitably, there is class E contamination produced by dust. This is often removed at later stages in the processing. Some people suggest that it is good workshop practice to remove dust from a printed circuit after each operation [2]. Oil and grease from machine parts, particularly from machines with steel tables, are also common causes of contamination.

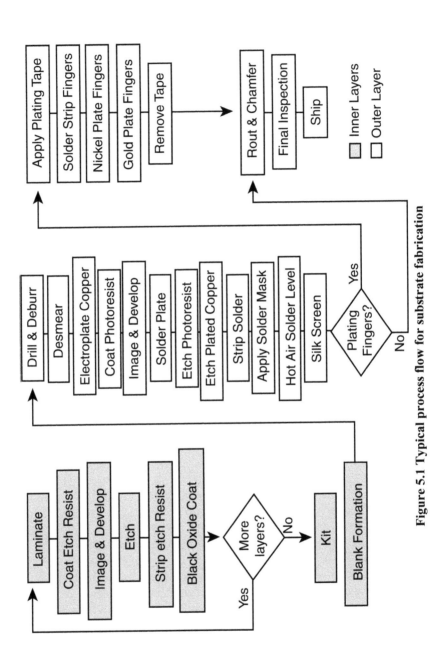

Figure 5.1 Typical process flow for substrate fabrication

5.2   Abrasive Brushing

A word of caution is needed here about abrasive brushing of metal plated (gold, silver, tin) surfaces. It is common in the industry to use electroless or immersion plating methodology, in order to leave a thin layer of metal. This is especially true for immersion plating. A layer of metal a few molecules thick may be all that is deposited in this process. Any attempts to use abrasives on these types of coatings can expose the boundary/base metal layer and create solderability problems later in the assembly process.

The most common mechanically induced contamination is produced by abrasive brushing processes, whether they are done by hand or machine. There are two types of contamination produced by brushing. The first is the implantation of abrasive particles into the metal regardless of the kind of abrasive used. Pumice and other pumice compounds, such as siliceous compounds, alumina, phosphates, oleic acid, and alkali compounds, can also be found on the substrate [4]. There is no way of avoiding this if a brushing process is used, and thus for critical applications it is frequently forbidden to brush and only chemical cleaning methods are used [13]. Such particles are the most common cause for dewetting during any hot-tinning or soldering process. It is easy to tin-lead plate or gold plate over these particles, producing an apparently perfect printed circuit. However, during soldering, either the tin-lead melts or the gold dissolves immediately in the solder; the intermetallics have to form on the copper surface underneath, which is contaminated with the abrasive particles [16]. One common way of reducing the quantity of implanted abrasives is to very lightly etch the surface with a persulfate bath to dissolve the copper around the particles. The particles are then free to fall or be washed off in a subsequent high-pressure rinsing process.

Of equal importance to the problem of abrasive implantation is smearing caused by the brushing process [15]. Smearing is produced when a brush picks up a contaminant and spreads it over the surface of the printed circuit. Typical examples of smearing occur in the following situations:

∞   Where bare copper is brushed; epoxy from bleed-through or pinholes can be smeared over the entire surface of the copper.

∞   A brush is used for cleaning a tin-lead plated board or a fused board and is contaminated with the tin-lead alloy; it is subsequently used for brushing copper, leaving small amounts of tin-lead on the copper.

∞   When a finished circuit board is brushed, epoxy can be picked up by the brush and later deposited on conductors.

∞   If any metal or metal oxide is picked up on a brush; it can be deposited on the epoxy between the conductors.

Ideally, a separate brushing machine should be provided for each operation; this machine can be fitted with the brushes most suited for that operation. This is extremely important when brushing the inner layers of a multi-layer board, because a small amount of contamination introduced by the brushing may be sufficient to degrade adhesion after pressing [15].

5.3   Chemical Pre-Treatment

The substrate is chemically prepared for the metallization application. Chemical decontamination typically consists of vapor degreasing with organic solvent and cleaning with mildly alkaline detergents, by either immersion or electrolytic action, to remove organic contaminants like oils and fats. Mild etching solutions are used to create a surface morphology on the copper cladding that is optimal for achieving bonds with the metallization [4]. Foreign metals, such as traces of aluminum from drill backup material and copper oxide, are also removed by these treatments. The cleaned and prepared surfaces are consequently sensitive to fingerprint contamination, airborne oxidants, and debris; so, good handling practices are very important.

If the substrate was inadequately prepared, then the hydrogen released during copper reduction can be trapped in the substrate holes at the interface with the deposited copper layer. The entrapped hydrogen causes out gassing problems and degradation of the plated hole [4]. The release of hydrogen is evident in the following reaction:

$$CuSO_4 + 2HCHO + 4NaOH = 2HCO_2Na + 2H_2O + Na_2SO_4 + H_2$$

Acidic pretreatments and acidic catalysts used for metallization can attack the glass treatment agents and encourage metallization to "creep" along the glass fiber into the laminate. Consequences are a degradation of the dielectric and the promotion of copper growth between biased holes with time, temperature, and humidity. This failure mode is referred to as conductive anodic filament growth (CAF). Damage done to the laminates during drilling operations, micro-delamination, uneven hole walls, broken glass fibers, create an area around these holes where CAF can more easily form. Experiments have shown this area to be about .25mm (.010 inches) wide.

5.4   Imaging

Plating and etch resists can be either photosensitive compounds applied as solid films or liquids, or liquid screening polymeric materials. Photosensitive compounds are applied, exposed, and developed to produce the required image. When masking is completed after plating or etching, the photoresist is stripped with appropriate media. The compounds can be either positive or negative working types. Insufficient development of these resists will leave contaminating residues, which, if not removed, can reduce solderability and the circuit reliability. When contaminated bare copper is solder-coated, the residues will leave places that remain unsoldered. The results are bare spots, pinholes, or inclusions, covered over by solder bridging [7].

5.5 Electroplating

By definition, all electroplating is highly ionic, so rinsing after any plating operation should be thorough. A simple ionic contamination test is sufficient to monitor the efficiency of this rinsing. It is common practice to use conductivity cells to monitor the water in rinse chambers in the plating line, but these are designed to reduce water consumption and do not control the quality of the rinse. Rinse chambers should not be overloaded by a throughput of circuits too great for the water throughput. An economizer should cut off the water between any two successive loads; if it does not, the water throughput must be increased.

Excess brightening agents (from the solder plating process), oxide films, and handling soils impair the adhesion of the electrodeposited metal to the contaminated surface [4]. Cross-contamination of plating tanks due to poor rinsing can occur and degrade the quality of the electrodeposits. This can be caused by insufficient rinse times or by inadequate turnover of immersion rinse tanks. Fresh spray rinses that follow immersion rinses are most beneficial, especially during the metallization operation prior to entry in plating baths. The purity of the rinse water is crucial; tap water contains many impurities that interfere with electroplating [4].

Metals used as anode material for electroplating are also a potential source of contamination. Anode impurities may become co-deposited and result in a contaminated electrodeposit. Other potential contaminant sources are foreign metals from poorly designed or maintained metal racks used for plating, and impurities in the air, especially oil from compressors used for agitating the plating baths.

5.6 Electroless and Immersion Plating

Electroless plating uses no electrical currents to transfer metals. It is quicker and less expensive then the traditional electroplating method, but there are some drawbacks to this process. The primary problem is the metal is porous. When gold plating is used nickel is the most commonly used boundary metal to prevent copper from leaching into the gold. The chemical reaction that causes the plating to occur also creates an uneven surface. The rough surface of the nickel combined with the porosity of the gold results in nickel and occasionally even copper bleeding through to the surface. Galvanic reactions, oxidation, and corrosion can occur which causes subsequent soldering problems. There is a shelf life for substrates that use this type of plating and thus proper storage and cleaning is critical for these units. As mentioned before, immersion gold plating, also known as flash gold, is only several microinches thick and requires even tighter controls for storage and handling.

Another problem with electroless nickel plating is contamination and chemistry of the bath. The chemical bath for the electroless plating of nickel requires constant monitoring to prevent build-ups of contaminants and to insure the proper chemical balance. An intermetallic layer which forms during soldering will result in an embrittled solder joint, leaving a smooth black or dark surface. This is sometimes referred to as black pad (see Chapter 11). There are several studies underway to determine the optimum amount of control that should be used in order to limit these types of solder joint failures [89, 90, 91, 92, 93].

5.7   Etching

The chemical removal of copper can be accomplished with a variety of etching solutions; the type of resist determines what solution can be used. The chemical nature of copper requires strong oxidizing agents to dissolve it effectively. The oxidants must be strong enough to oxidize the copper, yet gentle enough to avoid damage to the substrate material or the etch resist. Common copper etchants are ferric chloride, sulfuric acid, ammonium persulfate, cupric chloride, hydrogen peroxide, and alkaline etchants [7]. Contaminations originating from these solutions are mainly etchant residue from the etchant itself and residue from the interaction of the copper and resist material. When left on the surface of the resist metal or the etched portion of the substrate, the residue will decrease solderability, deteriorate the corrosion resistance, cause vesication, or decrease the surface resistance of the substrate [4].

The precipitation of cuprammoniacal complexes after etching in alkaline etchants is an important contaminating residue. These etchants generally operate in a pH range of somewhere between 8.5 and 10 [7]. As soon as the pH drops below 8, complex chemicals based on mixtures of copper, ammonia, and anions become insoluble and can precipitate on the circuits. In a properly controlled process, this should never happen during the actual etching. However, as soon as the circuit passes from the etching chamber to the first rinsing chamber, the pH can drop from the working value down to 7, and no amount of subsequent rinsing will effectively remove the complexes from the printed circuits [7]. This problem can be partially avoided if an ammonium hydroxide solution at a pH of over 8 is used in the first rinse section of the etching machine. This will keep the complexes in solution and dilute them enough to greatly reduce the amount of complex precipitate. However, this only works if the ammonium hydroxide solution is changed very regularly so that no great amount of copper becomes available [7].

The cuprammoniacal complex is by itself nonionic, but can readily be converted to ionic contamination in the presence of atmospheric humidity and carbon dioxide.

In critical applications, adequate cleaning is essential, preferably with ionic contamination control following the etching.

5.8 Tinning and Fusing Operations

Among the most common causes of contamination during printed circuit manufacture are fusing and leveling. The discussion here applies primarily to infrared and immersion fusing and hot air leveling. In all the cases mentioned, a flux is necessary for the process.

The formulation of the flux varies according to the nature of the process, but all fluxes contain a certain number of common compounds. The worst of these, from the point of view of contamination, are compounds based on polyglycols. A glycol is a heavy molecular-weight alcohol with two hydroxyl groups. The simplest glycol is ethylene glycol, commonly used as a radiator anti-freeze in cars. Its chemical formula is $OH-CH_2-CH_2-OH$. In the same family as ethylene glycol is diethylene glycol, with the formula $OH-CH_2-CH_2-O-CH_2-CH_2-OH$. Also in the same family is triethylene glycol, in which yet another $O-CH_2-CH_2$ linkage is added in the middle. This prolongation of the molecule or the increase of molecular weight is called polymerization; there is theoretically no limit to the length of a molecule [6].

Polyethylene glycols of molecular weights of over 6000 are frequently used in the manufacture of fluxes for fusing and leveling [6]. They have several ideal characteristics; they resist the fusing or leveling temperature very well; they are thick, almost waxy, materials that stay where they are required; and they provide an excellent protective barrier against reoxidation of the metal. Unfortunately, the majority of polyglycols have two effects that are highly undesirable for manufacturing printed circuits. The first is that the extremely long molecules have a tendency to stand on end on any surface, but more particularly on a lipophilic organic surface, such as the bare epoxy of a printed circuit or a solder resist coating. The exposed ends of the molecules have an OH radical, which makes the surface highly hydrophilic and easily wetted with water [4]. However, no amount of water washing, even with mechanical aids such as brushing, will effectively remove the entire monolayer of polyglycol molecules, because the molecules are tightly packed together. Furthermore, if a free ion is imprisoned in the molecular structure, it can cause a considerable drop in insulation resistance. One source of such an ion is the active part of the flux. These polyglycols are very hygroscopic, so they also contribute to a drop of insulation resistance, even without the presence of other contaminating ions, under humid climatic conditions [4].

The other phenomenon produced by the presence of polyglycols is that they can chemically attack the surface of epoxy resins and penetrate, compounding the effects described in the previous paragraph [10].

For circuits that are required to operate with low leakage currents, fusing or leveling are highly questionable; in the most critical cases, these operations can be described as a cause for degradation. It is possible to reduce the effects of polyglycol molecules to a minimum by judicious decontamination methods, which will be described later. However, no cleaning method will completely eliminate the problem [8].

5.9 Solder Masking

Vesication of a coating can be caused by the presence of ionic or nonionic contaminants under a protective coating. It is therefore important that a printed circuit be thoroughly cleaned and dried before a solder resist of any type is applied [4].

In one type of solder mask, contamination material is deposited on the pads to be soldered, or even into the plated through-holes. This could be due to poor housekeeping on a silk screen, inadequate development with a dry or wet film photo mask, or a number of other causes. This contamination must be considered class G contamination; it is probable that the printed circuit will become an irrecoverable reject [4].

A prerequisite to permanent solder mask application is an absence of water-soluble and adhesion-impairing residues. Hence, adequate cleaning must be applied, and cleanliness must be verified before coating. Plating and etch residues, retained flux, and handling soils from poor practices can cause serious problems when overcoated with a permanent mask. Poor quality of the rinse water is another source of residues under the solder mask. The processes must be well controlled to preclude coating over contaminated areas [4].

5.10 Drying

Another common cause of contamination during printed circuit manufacture is the use of shop air for drying the circuit at some stage. Compressed air should never be used for drying, since no matter how well it is filtered, it will contain water, oil, and acids. If an air process is necessary for drying, it is preferable to use a common hair dryer or, if high-pressure blow-off is required, bottled nitrogen [8].

5.11 Handling and Packaging

Handling and packaging techniques for bare circuits should be examined carefully. The effectiveness of the packaging depends on two conditions: it should not in itself introduce contamination onto the printed circuit, and it should prevent exterior contamination from reaching the circuit [8]. One of the most popular packaging materials, polyethylene has proven to be the least effective, particularly if heat-sealed. Polyethylene fails on both conditions; the plasticizers used to render the polyethylene pliant evaporate from the plastic and deposit themselves onto whatever is contained within the package. In addition, polyethylene is quite porous.

Unplasticized polyolefine is a much better material than polyethylene; even better is polyolefine laminated with aluminum [8]. The triple-layered bags used for storing sterilized foods, although expensive, offer one of the best protections. These consist of an outer layer of polyethylene, an inner layer of polyolefine, and in between, a thin layer of aluminum. These bags can be heat-sealed without any deterioration of the printed circuits inside. The only disadvantage of these bags is that they are not transparent, so a good labeling system is essential. If a printed circuit manufacturer goes to the expense of providing such packaging, the assembly plant can use the same bags after inspection and before assembly.

5.12 Summary of Contaminants

Potential contaminants from the substrate fabrication process, including their consequences are summarized in Table 5.2. Some of the common by-products produced from the substrate manufacturing process are listed in Table 5.1 and classified according to Table 1 in the Introduction. A detailed diagram of these contaminants is presented in Figure 5.2.

Table 5.1 Classification of each by-product according to solubility and mode of removal of by-product

By-Product	Classes
Carbonized residues	EFG
Polymers	ABCDG
Silica	EFG
Metal	EFG
Oxides	DE
Mineral acids	C
Salts	C
Cuprammoniacal complexes	D
Organo-metallic salts	D
Polyacids	CDG
Inorganic acids	C
Organic acids	BCD
Nitrogenic compounds	CD
Compounds with N and S	DG
Silicones	G
Particles	EFG

Table 5.2 Summary of potential contaminates from substrate fabrication

Contaminant Sources	Contaminants	Consequences
Plant air and machinery	Dust, dirt, oil, grease, epoxy smear, foreign metals	Poor solderability, degradation of dielectric properties, poor metal adhesion
Abrasive brushing	Pumice, siliceous compounds, alumina, phosphates, oleic acid, alkali compounds	Dewetting, poor solderability, pinholes, epoxy bleed-through
Chemical pre-treatment	Solvent residues, fingerprints, entrapped hydrogen	Poor adhesion, delamination, degradation of dielectric properties, out gassing
Imaging materials	Etch resist residues, organic residues, metallic residues	Poor solderability, bare spots, pinholes, solder bridging
Electroplating materials	Excess brightening agents, organic residues, oxide films, water contaminants, alkaline solution residues, anode metals, foreign metals, plating bath residues	Poor metal adhesion, poor electrodeposit quality, nodular plating, out gassing, PTH and via cracks
Etchants	Ferric chloride, sulfuric acids, ammonium persulfate, cupric chloride, alkalines	Decrease solderability, deteriorate corrosion, resistance, vesication, decrease substrate surface resistance
Tinning and fusing	Flux residues, polyglycols, polyethylene glycols	Degrade dielectric properties, current leakage
Solder masking	Trapped residues and particles, moisture, air	Irremovable contaminants, out gassing, poor solderability
Drying	Water, oils, acids	Poor solderability, decrease substrate surface resistance
Handling and packaging	Fingerprints, dust, dirt, dander, airborne particles, moisture, plasticizers	Poor solderability, decrease substrate surface resistance, deteriorate corrosion resistance

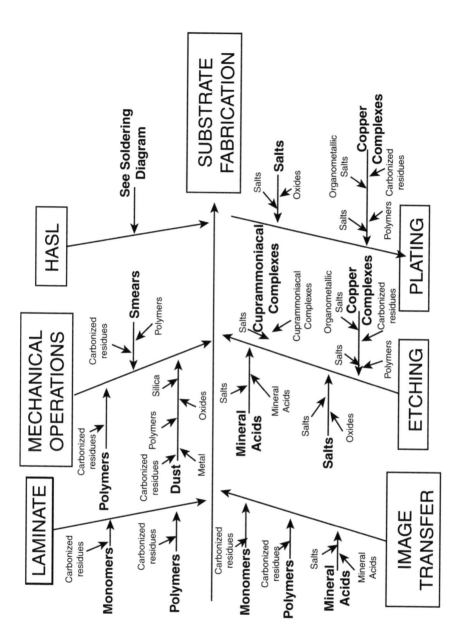

Figure 5.2 Detailed diagram for tracing contaminants during substrate fabrication

CHAPTER 6   MEASURING CONTAMINANTS IN SUBSTRATES

According to IPC standards and specifications, the first formal cleanliness testing is at the printed circuit level. This is the first time during the manufacture of a substrate that ionic and nonionic contaminants are measured and tested. The measurement of the different contaminants is discussed in the assembly section of this document, and the procedures used at the assembly level are also used at the substrate level.

According to IPC-A-600, "Acceptability of Printed Boards," the following general rules minimize surface contaminants when handling printed boards [72]:

- Workstations should be kept clean and neat.
- There should be no eating, drinking, or use of tobacco products at the workstation or other activities that are likely to cause contamination of the board surfaces.
- Hand creams and lotions containing silicone should not be used since they can result in solderability and other processing problems. Specially formulated lotions are available.
- Handling of boards by their edges is desirable.
- Lint-free cotton or disposable plastic gloves should be used when handling bare boards. Gloves should be changed frequently as dirty gloves cause contamination problems.
- Unless special racks are provided, stacking boards without interleaving protection should be avoided.

The IPC-A-600 standard is used to determine whether contaminants are organic or inorganic, ionizable or nonionizable, and recognizes the six most common contaminants found on printed circuits [72]:

- Flux residues
- Particulate matter
- Chemical salt residues
- Fingerprints
- Corrosion (oxides)
- White residues

Ionic contamination is specified to be no more than 1.56 $\mu g/cm^2$ of NaCl or equivalent. For organic contamination, the IPC-6012 standard, "Qualification and Performance Specification for Rigid Printed Boards," states, "non-coated printed boards shall be tested in accordance to IPC-TM-650, methods 2.3.38 or 2.3.39. Any positive identification or organic contamination shall constitute a failure." The requirements for the maximum allowable level of rosin flux residues are 200 $\mu g/cm^2$ for class 1 assemblies, 100 $\mu g/cm^2$ for class 2, and 40 $\mu g/cm^2$ for class 3. The classifications are defined in J-STD-001, "Requirements for Soldered Electrical and Electronic Assemblies" as:

Class 1: *General Electronic Products.* Includes consumer products, some computer and computer peripherals, and hardware suitable for applications for which the major requirement is function ability of the completed assembly.

Class 2: *Dedicated Service Electronic Products.* Includes communications equipment, sophisticated business machines, and instruments requiring high performance and extended life, for which uninterrupted service is desired but not critical. Typically the end-use environment would not cause failures.

Class 3: *High-Performance Electronic Products.* Includes equipment for commercial and military products for which continued performance or performance-on-demand is critical. Equipment downtime cannot be tolerated, end-use environment may be uncommonly harsh, and the equipment must function when required, such as life-support systems and critical weapons systems.

The completed boards should also not support fungus growth, according to test method 2.6.1. The ionic and organic contamination test methods in accordance with IPC-TM-650 are described in Table 6.1.

Table 6.1 Test Methods for contamination testing according to IPC-TM-650

Test Method	Test Subject	Comments
2.3.25	Resistivity of solvent extract	Determines the presence of ionic contaminants via a drop in electrical resistivity.
2.3.26	Ionizable detection of surface contaminants (dynamic method)	The soil must be soluble in water, alcohol, or some mixture of both. The results are not species-specific, but ranks them from highest mobility to lowest. Comparable to an ionograph.
2.3.27	Cleanliness test - residual rosin	Quantifies the residual rosin left from soldering operations.
2.3.38	Surface organic contaminant detection test (in-house method)	Is not species-specific.
2.3.39	Surface organic contaminant identification test (infrared analytical method)	A continuation of Test Method 2.3.38 to identify the organic contaminants and put them into classes.
2.6.1	Fungus resistance printed wiring materials	Tests to determine the resistance of materials to fungi, and if such material is adversely affected under conditions favorable for their development (high humidity, warm atmosphere, and presence of inorganic salts).
2.6.3	Moisture and insulation resistance, rigid, rigid/flex, and flex printed wiring boards	Determines visual degradation and electrical insulation resistance due to deleterious effects of high humidity and heat conditions of tropical environments.
2.6.3.3	Moisture and surface insulation resistance, fluxes	Characterizes fluxes by determining the degradation of electrical insulation resistance of PWBs due to high humidity and high temperature conditions.

CHAPTER 7   QUALITY AND RELIABILITY OF SUBSTRATES

Like the defects created during laminate processing, more defects can occur during substrate manufacture. Given the similarity between laminates and substrates, some of the defects discussed in the laminate section also develop during substrate manufacture. Sometimes, defects are not detected at the laminate stage, but the abundance of steps required to produce a multilayered substrate precipitates some defects, and they become observable. These defects can impact the assembly process, because they can grow, reducing the reliability of the finished product. Some of the defects observed at the substrate level are discussed in this chapter.

7.1   Nonwetting and Dewetting

Nonwetting and dewetting (Figure 7.1) occurs at the substrate level as well as the assembly level, but usually for the same reason: the metals are contaminated with a substance that will not allow the solder to wet. These defects can lead to electrical opens if the two metals are not soldered correctly, or to electrical shorts if the solder bridges to other conductors.

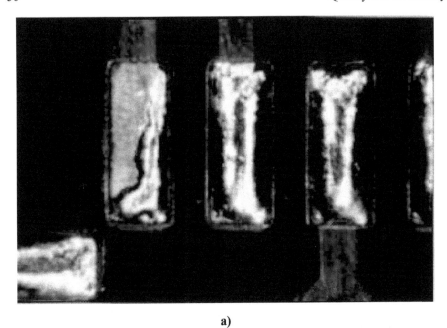

a)

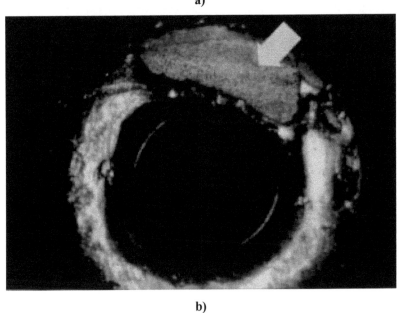

b)

Figure 7.1 a) Non wetting solder on a bond pad,
b) dewetting solder on a plated through hole [72]. With permission from the
Institute for Interconnecting and Pacakaging Electronic Circuits.

7.2 Plating Voids

Voids, due to contaminants on the surface, can occur while plating through-holes or bond pads (Figure 7.2). These voids can lead to poor electrical conductivity, leading to an open circuit.

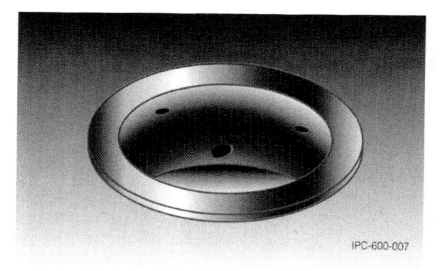

Figure 7.2 Plating voids on a copper-plated through-hole [72]. With permission from the Institute for Interconnecting and Packaging Electronic Circuits.

7.3 Haloing

Haloing (Figure7.3) is mechanically induced fracturing or delaminating on or below the surface of the base material. It is usually observable as a light area around the holes, other machined areas, or both. This defect can be caused by contamination that would prevent a good bond between the laminate material and the copper-plated hole.

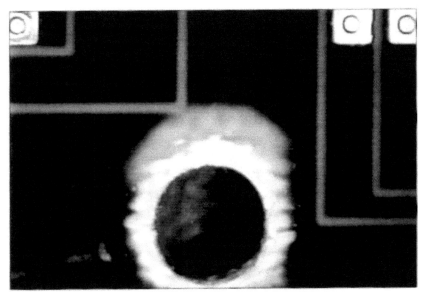

Figure 7.3 Haloing of a copper-plated through-hole [72]. With permission from the Institute for Interconnecting and Packaging Electronic Circuits.

7.4 Lifted Lands

Conductor lands can be separated from the base material due to contaminants that causes poor adhesion between the two materials (Figure 7.4). This condition can cause the whole land to become detached, causing a loss of signal transmission.

Figure 7.4 A copper land detached from the base material [72]. With permission from the Institute for Interconnecting and Packaging Electronic Circuits.

7.5 Resist Deadhesion

Some solder resists are used as a permanent coating on a substrate to protect the surface from other contaminants, corrosion, mechanical damage, and solder bridging. However, if the resist does not adhere to the surface, its protection is minimized or eliminated (Figure 7.5).

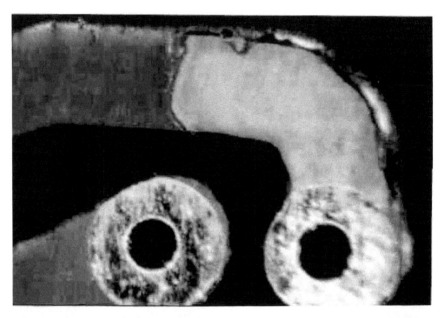

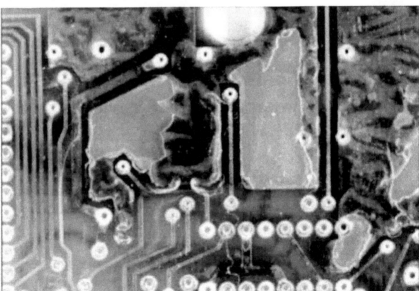

Figure 7.5 Photographs of the flaking of solder resist covering
a copper-traced substrate [72]. With permission from the Institute for
Interconnecting and Packaging Electronic Circuits.

7.6   Resist Blistering and Waving

The solder resist can blister due to entrapped air, moisture, or other contaminants, as well as appear wavy after it is applied (Figure 7.6). The wavy appearance can be caused by the contamination of the entire surface being contaminated, and can prevent adhesion of the coating. Open areas between the coating and base material provide a path for moisture or other ionic contaminants, creating a conductive path between conductors, leading to electromigration and a short.

7.7   Vesication

Vesication is another form of blistering. If, for example, a layer of solder resist or conformal coating covers a crystal of salt, it might be assumed that no harm will result. However, all organic coatings are somewhat porous, and the protective coating is a semi-porous membrane. Under humid conditions, osmosis will occur and the salt crystal will attract humidity through the coating. The salt may eventually dissolve partially or entirely in the liquid that accumulates, and an equilibrium state will be reached when the internal osmotic pressure equals the vapor pressure on the outside. This may be several g/cm^2, and sufficient to detach the coating from the substrate. Under a microscope, a vesicle thus formed may sometimes be identified, particularly if the salt particle is large. Most frequently, vesication manifests itself as a milky appearance under the coating or in it. Any substance, ionic or not, which is at all hygroscopic may cause vesication. If the vesication is due to ionic contamination or nonionic with carbon dioxide absorption, leakage currents can occur under the protective coating.

It is essential that printed circuits be properly cleaned with suitable solvents, followed by distilled and deionized water, and thoroughly dried before any protective coating is applied at any stage of manufacture. The more critical the application, the more important this becomes. It is good practice to perform an ionic contamination test before applying any kind of protective coating, whether the circuit is with or without components. Of course, this test will only detect the presence of ionic contamination, but it is rare for nonionic contamination to exist alone.

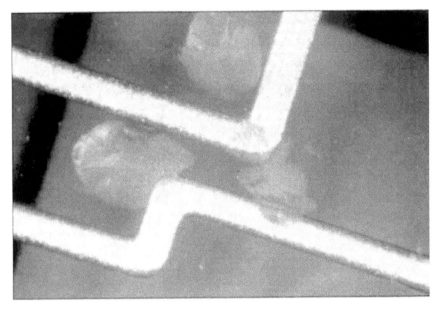

a)

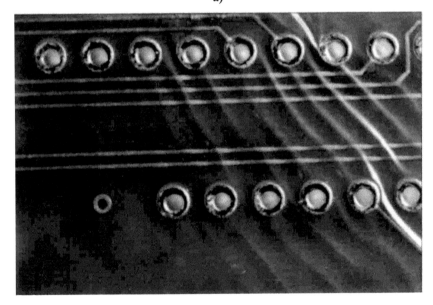

b)

**Figure 7.6 Photographs showing blistering(a) and waving
(b) of solder resist [72]. With permission from the Institute for Interconnecting
and Packaging Electronic Circuits.**

7.8   Soda Strawing

Soda strawing describes a long tubular void along the edges of conductive patterns where the solder resist is not bonded to the base material or to the edge of the conductor (Figure 7.7). Tin/lead fusing fluxes, fusing oils, solder fluxes, cleaning agents, or volatile materials may be trapped in the soda straw void, causing detrimental effects similar to other solder resist defects.

Figure 7.7 Illustration of soda strawing of solder resist over three adjacent conductor traces [72]. With permission from the Institute for Interconnecting and Packaging Electronic Circuits.

7.9   Blistering

A printed circuit with a layer of silicone product over it, to take an extreme example, will prevent any solder resist or conformal coating from adhering and will flake rapidly. Fortunately, silicones are rarely used, but many other substances can also reduce adhesion, less dramatically. If this happens over the whole surface, then flaking will occur; if it is present only locally, then blistering might result.

7.10 Delamination

Delamination of assembled multilayers is principally associated with contamination, or a lack of suitable surface preparation. During the pressing cycle, the prepreg has to be made to adhere successfully both to the copper of the inner layers and to the epoxy resin where there is no copper. The surface chemistry can be quite complicated.

Where the copper to no copper ratio is very low, it is often sufficient to pumice the copper to provide sufficient adhesion- obviously under controlled conditions. In other cases, one or more of the following processes may be necessary: brushing using a special adhesive brush, chemical cleaning using a proprietary cleaner, chemical oxidation of the copper, or special decontamination. Oxidizing baths are highly alkaline and the residues are very difficult to remove. Ionic contamination control is desirable before laminating.

The surface treatment that is ideal for copper may be less so for the epoxy resin. Without copper etched from it, a structured surface must be specified when ordering the laminate. If etched, the surface will promote good adhesion, provided that it is clean and dry. Several hours in a vacuum oven at 60°C may be desirable to ensure maximum evaporation of water and solvents. On suitably decontaminated surfaces, no blistering should occur after the circuit has been submitted to immersion in a solder bath for 40 to 60 seconds at 250°C.

7.11 Particulate Contamination

Particulate contamination can cause problems. For example, the danger of a metallic sliver is obvious. However, other solid particles have been known to cause malfunctions of electronic circuits, particularly when hair or fibers are bridging two conductors. If these hairs or fibers are fixed in position by the encapsulation of a conformal coating or a solder mask, a leakage current can arise under humid conditions. Whether this seriously affects the function of the circuit depends on its impedance and design.

Airborne particulates range in size from .002 to over 10 microns. In a typical polluted urban environment, the total amount of suspended particles can be as high as 10^8 per cm^3. A large portion of these particles is highly ionic (up to 50%). According to a Bell Laboratory report, the total dry deposition rate for particles in a standard outdoor urban environment is up to 2×10^9 per ft^2 per second (for .01 to 2 micron sizes). Indoor deposition depends on numerous factors, including the quality of filtering systems, leakage of outdoor air, and indoor generated contaminates. The deposition rates can vary from 20% to as high as 80% of the outdoor rates [84]. Periodic monitoring of fine particle deposition is advised to prevent production or reliability problems, as many of these particles are very hygroscopic [102].

CHAPTER 8   CONTROL METHODS IN SUBSTRATE MANUFACTURING

Control methods discussed regarding laminates also apply to substrates and references to cleaning, shipping, handling, and packaging should be noted here as well as in subsequent chapters. The details of such topics are discussed primarily in the assembly section. Likewise, the discussion of protection methods is confined to the assembly section.

8.1   Manufacturing Process Assessment

The assessment of the manufacturing process through proper control items can help to identify if the process is releasing contaminants into the product. Currently there are quite a few standards available for substrates. Most notable for manufacturing process assessment is IPC-2524 "PWB Fabrication Data Quality Rating System." This document is not specific to contamination guidelines, however; so the rest of this chapter presents some recommendations for the control of contamination in the substrate manufacturing process.

8.1.1 Quality Assurance through Certification

In order to monitor processes that may allow contamination, quality assurance through certification or external audits should be considered. This takes the viewpoint of say what you're going to do, do what you say, and document what you do. At the minimum this requires a process control manual. It is preferable to have the PCB manufacturer be certified by ISO, as this incorporates external audits of the quality system.

8.1.2 Statistical Process Control

The measurement of critical parameters within the manufacturing process can target when a process or group of processes are not performing acceptably. This can also help to identify potential contamination sources.

One recommendation is the use of ion chromatography for statistical process control (SPC), considering the following criteria for guidance:

- Use of ion chromatography BEFORE application of solder mask
- Use of ion chromatography BEFORE inspection and test
- Test should be performed at the beginning and end of each shift
- Results should not be averaged
- Chloride levels should not exceed 2 $\mu g/in^2$
- Bromide levels should not exceed 10 $\mu g/in^2$
- SPC results should be easily accessible for operators (preferably posted)
- Evidence of segregation of product exposed to out-of-specification process
- Evidence of action plan for process out of control

8.1.3 Cleaning

Cleaning methods vary from manufacturer to manufacturer. It should be noted that incomplete removal of cleaning solvents and/or residues can leave detrimental contaminants on the laminate, reducing the reliability of the finished assembly. This is especially important where etch chemistries are concerned.

8.1.4 Best Practices

Best practices are usually not required for acceptable products, but are strongly recommended for quality product. With this in mind, a list of best practices to monitor contamination for substrate manufacturing is in section 8.4.

8.2   Control Monitoring through Testing

Using tests for control monitoring at key points throughout the substrate manufacturing process is recommended. Some common tests have been listed here. This list is not meant to be exhaustive, but is meant to provide examples of some tests that can be conducted to monitor the manufacturing process.

8.2.1 Visual Inspection

Visual inspection is a manual approach in that it makes use of people, good lighting, and some type of training on what is acceptable and what is not. Usually a comparison to a known good product or the artwork is made. If the inspection person has seen the board often, he or she becomes more skilled at finding faults and looking in likely locations for faults. Visual inspection may be appropriate for the aesthetic or visual appearance of the board. Aesthetics can be checked by a human. There is no question that what appears as poor solder mask application or scratches on a product may be a latent defect that will later lead to failure of what is presently a good board. This falls outside the realm of electrical test.

8.2.2 Automated Optical Inspection

Other than the human eye, there are computer-based visual inspection methods, referred to as automatic optical inspection or AOI. AOI systems automatically scan printed circuit panels to detect circuit pattern defects such as shorts, opens, lead reductions and projections. AOI equipment checks the board or its innerlayers against dimensional parameters that have been programmed into it. These can be generally accepted or design rule-based parameters, or they can be programmed to recognize a window of acceptable dimensions for features on the board. Like manual visual inspection, faults found with this method can imply that there may be an impact on the board's functionality, but the board's functionality and interconnect are not actually tested. Distinctions between the aesthetics of the board's features and its fitness for use are difficult to differentiate. As such AOI is typically used for innerlayer testing but does not serve as a final test.

8.2.3 Plated Through Hole Reliability Test

In plated through hole (PTH) reliability tests, the PTH is subjected to a temperature change from -65°C to 125°C or from 25°C to 260°C. For these reliability tests, micro-sectioning is usually employed for the evaluation of the PTH. An acceleration factor is calculated to determine expected life in the field. Some typical PTH reliability tests employed are:

- MIL-P-55110E Thermal Shock Test
- MIL-STD-202 Temperature Cycling Test
- IBM 6X Solder Shock Test
- APD-Oil Shock Test
- IEC Oil Shock Test
- IPC TM-650-2.6.6B– Temperature Cycling, Printed Wiring Board; 12/87
- IPC TM-650-2.6.7A– Thermal Shock and Continuity, Printed Board; 8/97

- IPC TM-650– 2.6.7.2A-Thermal Shock, Continuity and Microsection, Printed Board; 8/97
- IPC TM-650-2.6.8D– Thermal Stress, Plated Through-Holes; 3/98

8.2.4 Insulation Reliability Testing

The primary function of a printed circuit board is to provide either low impedance or high impedance where needed, thus forming the basis for electrical interconnection. Reliability problems are encountered when there is a degradation of either (opens or shorts). The typical insulation resistance reliability test employed is IPC TM-650-2.6.3E– Moisture and Insulation Resistance, Printed Boards; 8/97.

8.2.5 Electrical Continuity Testing

Electrical test of most circuit boards requires preparation of a test fixture that is unique to each specific board design. The fixture interfaces the test system to a field of test points arranged to probe the features of interest on a specific product. Various types of fixtures can be used, depending upon the product, and the type of test system employed. The cost of preparing this test fixture is ultimately passed on to the customer. The total number of points contacted, requirements for simultaneous dual–side probing, and the pitch (spacing) of the test points, each affects fixture cost. Very closely spaced targets of small size require the use of more expensive probing techniques. The fixture design process includes decisions as to which substrate features should have assigned test points. These decisions can affect whether the substrate actually receives a valid 100% test.

The following key questions should be considered during the preparations of specifications for bare–board test:

- Are your board test requirements clearly stated?
- Which board features must be tested?
- Are your electrical test criteria appropriate for the intended end use of the board?
- What source data will be used for development of the electrical test program and fixture?

Electrical test parameters usually include the following:

- Continuity Resistance Pass/Fail Threshold - Expected connections, which exhibit resistance in excess of the value set for the continuity resistance test Pass/Fail threshold, will FAIL the test, and should be reported as OPEN.
- Maximum Continuity Test Current - Since the conductors on modern circuit boards are considerably finer, the maximum test current that can be delivered to the board should be considered, without risk of damage. A few systems with stimulus levels of up to one ampere are still in use. Therefore, maximum amount of current that may be passed through the board during continuity test should be specified, without limiting the

value excessively. This will permit use of the widest range of test systems, while protecting the product from damage.

- Isolation Resistance Pass/Fail Threshold - This test should verify that each network is well isolated from the rest of the board. The test system will ground all networks except the network under test. The test system will then measure the resistance between the network under test and the ground representing all other networks. The system detects the parallel combination of all leakage resistance, from the tested network to the balance of the board. If the measured resistance is less than the isolation test resistance Pass/Fail threshold, the tested network will FAIL.

- Voltage applied to passing networks during Isolation Test – During the isolation test of each network, a small amount of current is injected into the network. If the network is well isolated, it is elevated in voltage by this current. The network will act as a capacitor, and the arriving current will steadily charge it to higher voltage levels. In this case the voltage of the network under test will ramp up to a value determined by a voltage limiting circuit in the test system. On the other hand, if the network leaks significant current to other networks, then the test current may be unable to charge it up to the voltage limit. This is the case when the isolation test fails. Higher voltages are useful in assuring accuracy for the higher isolation test resistance Pass/Fail thresholds. A given leakage resistance will leak more current with a higher applied voltage, and is more easily detected. High voltage can also encourage the breakdown of ionic contaminants, increasing the leakage current flow and helping to ensure their detection during test. These are the primary reasons for specifying the open–circuit test voltage during the isolation test of the product. Of course, the voltage should not be so high that there is any risk of damaging the product.

- Isolation Test Method – The "flying–probe" test systems are unable to contact all networks simultaneously, so they can't measure the total leakage from a given network to the entire balance of the board in a single measurement. They are limited to measuring the isolation resistance between specific network pairs. The flying–probe systems can detect hard shorts and concentrated isolation resistance failures, but they are less sensitive to distributed contamination type defects. In view of this limit then, for speed reasons, they are often restricted to measurement of nets physically adjacent to the network in question. If you're concerned with high–impedance/high–voltage isolation test results, you should consider the limitations of this type test system.

Any special requirements related to components embedded within the substrate, characteristic impedance (Zo), or other special features should also be considered. (A Time Domain Reflectometer or Network Analyzer would typically be used for this purpose.)

8.3 Physics of Failure Based Root Cause Analysis

Root cause analysis procedures should be in place for when a noncomformity is found in the product. These procedures must be based on physics of failure (POF), so the failure mechanism is correctly identified and the problem corrected.

8.3.1 Material Traceability

Materials should be marked in such a way as to know manufacture date, time, and lot number. Contamination is lot sensitive in many cases, making materials traceability an extremely important part of the root cause analysis. Materials traceability is also dependent on the vendors used for all the supplies necessary to make a substrate. Some best practices for these vendors are listed in section 8.4.

8.3.2 Corrective and Preventive Actions

Once the nonconformity is found and traced to the root cause, a procedure for corrective actions should be in place so there are guidelines already in place to effectively correct the problem. These procedures can directly relate to guidelines for preventive actions, so once a nonconformity occurs and is corrected, it will not happen again.

Corrective actions (CAs) are activities designed to prevent the recurrence of a failure cause. Corrective actions are generated from material review board action, customer audit or internal audit. Formal corrective actions are developed during brainstorming sessions, quality meetings, design reviews, and other structured improvement activities. Corrective actions may be given based on highest failure cost, customer complaints, unacceptable quality levels of raw materials or internal quality audit findings.

The corrective action request may be assigned by the corrective action coordination, the corrective action team or by the quality person. When appropriate, the corrective action should be assigned to an investigative team. This team (or individual) uses the method for identifying the root cause and solving the problem as detailed in the corrective action procedure. All actions should be recorded and turned into the quality administration department. All corrective action responses should be reviewed for accuracy, effectiveness, and completeness by the corrective action team, chaired by the Quality Administration Manager.

Opportunities for improvement that would eliminate potential causes of nonconformity's shall be identified as "Preventative Actions." Sources of information may come from processes and work operations that affect product quality, audit results, customer complaints or actions as a result of the management review.

8.3.3 Documentation

A contamination database which records possible sources of contamination and the proper methods for correction should be part of the company's documents. This allows for the experience of employees to be passed to those with less experience.

Process Change Notification. Change control/notification systems are intended to notify customers about the changes in the product or process before the changes are implemented. Substrate manufacturer's notification policy should provide support for conducting the necessary changes and notifying other persons to whom these changes may be of interest.

The design data management system should support information that exists in small design teams. The design system program provides data storage and retrieval, data transfer, security, version control, release control, change notification, historical tracing, error reporting, and archiving. Engineering orders, requests for waivers or deviations, and engineering change proposals should be reviewed to determine if they are correct, complete, and in compliance with standards and needs. The review assures compliance with contract requirements for proposing, approving, and implementing engineering changes, including the monitoring of changes requiring approval by on-site and off-site customers, as well as those changes that do not require customer approval.

Effective points for change incorporation should be established for changes that have been approved, and deliverable documents are provided to the customer based upon data acquisition requirements of the contract. The quality group should upgrade inspection and test instructions, manufacturing travelers, and routing tickets as required by the approved change. Approved production and assembly changes should be monitored at the designated effective points, and corrective action should be taken where discrepancies are found. The change should be immediately distributed to the planning department and incorporated into the working schedule, for calculation of any schedule impact.

8.4   Supply Chain Assessment

This section provides tables of questions that are considered best practice to ask when assessing the supply chain of a substrate manufacturer. The tables are meant to serve as examples of how a supply chain assessment can reveal possible sources of contamination. It is a best practice to assess all vendors. Table 8.1 provides an example assessment for drilling and back-up materials vendors. Table 8.2 is an example assessment of plating materials and oxide coating vendors. Table 8.3 is an example assessment of solder mask vendors.

Table 8.1 Drilling entry and back-up materials vendor assessment

DEFECT/ FAILURE MECHANISM	POLICY/PROCEDURE	YES	NO
Organic contamination	Does the entry material supplier document the organic resin composition for every purchase lot?		
Insufficient or excessive hardness of entry material	Does the entry and backup material suppliers document the hardness of every purchase lot?		
Poor drillability	Does the PCB manufacturer re-optimize the drilling process with every change in entry material supplier?		
Low T_g resin of backup material	Does the backup material supplier measure T_g on each purchase lot?		
Deformation, non-uniform thickness and density variation of backup material	Does the backup material supplier perform the bow and twist measurement for each lot?		
	Does the backup material supplier provide thickness and density measurement data?		

Table 8.2 Plating materials and oxide coatings vendor assessment

DEFECT/FAILURE MECHANISM	POLICY/PROCEDURE	YES	NO
Composition out of specification	Does the plating materials supplier document the composition for every purchase lot?		
	Does the oxide coatings supplier document the composition for every purchase lot?		
Mechanical properties of plating materials below specifications	Does the plating materials supplier perform ductility and tensile strength testing on each batch of plating material?		
Degradation of plating materials and oxide materials	Does substrate manufacturer store plating materials and oxide coatings based on manufacturer's recommendations?		
	Does substrate manufacturer have policies and procedures to prevent use of expired plating materials and oxide coatings?		

Table 8.3 Solder mask vendor assessment

DEFECT/FAILURE MECHANISM	POLICY/PROCEDURE	YES	NO
Composition out of specification	Does the solder mask supplier document the composition for every purchase lot?		
Thickness below specifications	Does the solder mask supplier document the viscosity and thixotropic index for every purchase lot?		
	Does substrate manufacturer follow solder mask supplier's recommendations in regards to application?		
Color/Reflectivity out of specification	Does the solder mask supplier document the color and reflectivity for every purchase lot?		
Excessive porosity	Does the substrate manufacturer benchmark solder mask porosity by performing surface insulation resistance (SIR) testing with every change in solder mask material or solder mask manufacturer?		
Solder mask not fully cured	Does the substrate manufacturer follow solder mask supplier's recommendations in regards to cure?		
Incompatibility with plating/ soldering/cleaning processes	Does the substrate manufacturer perform qualification tests with every change in solder mask material or solder mask manufacturer?		
Degradation of solder mask material	Does substrate manufacturer store solder mask materials based on manufacturer's recommendations?		
	Does substrate manufacturer have policies and procedures to prevent use of expired solder mask materials?		

8.5 Shipping, Handling, and Storage

Handling assessment issues in substrate manufacturing covers board handling in innerlayer processing, multilayer lamination, and image transfer, as well as during receiving and packing. Storage issue covers the storage of materials (laminates) used in manufacturing and also storage of the substrates in process and before shipping. Shipping issue considers how the substrate is packaged and traceability issues. The most important issue in handling, storage, and shipping is to provide the methods to avoid damage to substrate and its supply chain material from moisture absorption and to avoid the use of material whose shelf life is finished.

The substrate manufacturer handling, storage, and shipping issue policies and procedures should be designed to ensure that systems are in place to avoid the occurrence of defects or physical mechanisms that can cause failure during board manufacture assembly because of wrong handling, storage, and shipping. For the manufacturing process the big picture should be in consideration, i.e., from the incoming door to the outgoing door, and everything in between, not just the manufacturing floor, and this includes also how the finished product is handled, packed, and shipped.

PART 3

PRINTED WIRING BOARD ASSEMBLY

CHAPTER 9   CONTAMINANT SOURCES IN PRINTED WIRING
BOARD ASSEMBLY

As with printed circuit manufacture, nearly every operation that a printed circuit undergoes from the time it reaches the assembly plant to the time it leaves can introduce contamination. This chapter examines the various assembly operations in the order in which they generally occur, and the kinds of contamination that can arise.

9.1 Receiving and Handling

The first thing that generally happens to printed circuits when they reach an assembly plant is that the carefully wrapped circuits are unwrapped for inspection. At best, they will be subject to fingerprints, and at worst to a large number of other contaminants.

9.1.1 Fingerprints

Fingerprints are extremely complex chemically and can vary from person to person. They consist mostly of oils, water, and salt, but there are also amino acids and urea [49]. To make things worse, hands are frequently contaminated with other substances sometimes deliberately, such as hand lotions and creams, and sometimes accidentally, such as improper rinsing after hand washing. Therefore, in addition to purely organic products exuded by the skin, hands may carry soap, heavy alcohols, fatty acids, vegetable and animal fats, petroleum derivatives, or even silicones [49]. From certain hand cleaning products, abrasives can be introduced, and from cloth or paper towels, fingers can deposit lint onto the printed circuit. If possible, all people handling printed circuits or any other electronic component should be required to be equipped with suitable protection to prevent fingerprinting [49].

Fingerprints containing organic acids react with different metal ions, including copper, silver, tin, and lead ions. They affect surface resistivity and promote dendrite formation [51].

A detailed diagram of contaminants from fingerprints is shown in Figure 9.1. For classification of the contaminants, refer to Table 1 (pg. x) for each by-product.

9.1.2 Gloves

It might seem that rubber gloves would be the ideal answer, but rubber gloves themselves are usually saturated with some kind of plasticizer or lubricant, not to mention talcum powder [48]. Clean woven gloves appear to be a better idea, although there is an increased risk of lint.

Light, knitted white cotton gloves usually present the best compromise when handling electronic assemblies [48]. These gloves should be considered throwaway items, and must therefore be relatively low-cost. In order to reduce lint formation and possible contamination, it is strongly recommended that gloves be washed before use. In addition, the operator's hand should be carefully washed and rinsed free of any skin lotion or cream and perfectly dry before the gloves are put on.

9.1.3 Food, Drink, and Tobacco

Where electronic assemblies are handled or assembled eating, drinking, and smoking should be strictly forbidden. Most food and drink are acidic (some soft drinks have a pH of about 2) [47], and it is not enough to keep the food on the other side of the room. Simply peeling an orange, for example, can spray a fine mist of acidic oil particles over a distance of 3 to 4 meters [47].

9.1.4 Airborne Particles

Airborne contaminants or pollutants exist in three chemical stages. The first is reactive gaseous agents (NOx, SOx, O_3, H_2O_2 , NH_3 , H_2S, plus VOCs [Volatile Organic Compounds]). The next stage is suspended inorganic and organic acids (H_2SO_4, HNO_3, HCl, $HCOOH$). The final stage is suspended submicron hygroscopic ionic particles (NH_4HSO_4, $NH_4 \cdot NO_3$, $[NH_4]_2SO_4$) plus numerous others. Additionally, there are many other inorganic compounds and metals, such as sodium chloride, iron, aluminum, magnesium, calcium, and potassium, suspended in the air.

Unfortunately, all three stages can be present at any given time. Which stage is predominant depends on the time of year, temperature, humidity, local weather conditions, effectiveness of environmental control methods, and the length of time the circuitry has been exposed. All three stages generally require the presence of moisture to cause the chemical reactions necessary to degrade circuit reliability.

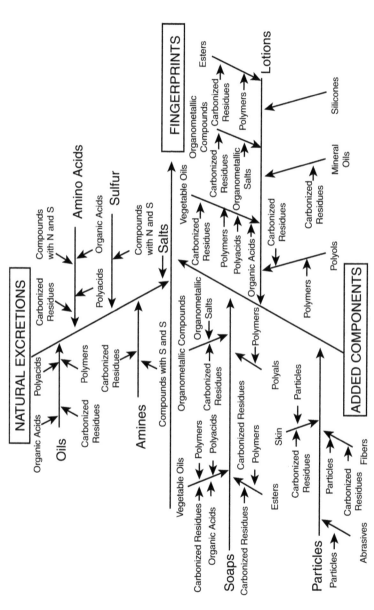

Figure 9.1 Detailed diagram of contaminants from fingerprints

Reactive gases usually have a relative short life expectancy in the atmosphere. They will last for a few minutes to several days depending on conditions and type of gas. Many gases combine to form other airborne compounds as well as reacting with or being adsorbed/absorbed by surfaces.

Table 9.1 Concentration of selected gaseous air constituents (ppb) in the United States [4, 150]

Gas	Outdoor (ppb)	Indoor (ppb)
O_3	4-42 (250)	3-30 (125)
SO_2	1-65 (150)	0.3-14 (50)
H_2S	0.7-24 (40)	0.1-0.7 (40)
NO_2	9-78 (250)	1-29 (200)
HNO_3	1-10 (50)	3 (15)
NH_3	7-16 (50)	13-259 (500)
$HCl + Cl_2$	<0.005-0.08* (6)	0.001-0.005 (5)

*Corresponding to 5 wt% HCl

Table 9.1 shows a generalized arithmetic mean level (with 95th percentile levels in parentheses) of the amount of gaseous pollutants in the air for the United States. Other parts of the world, notably China/Asia, East Central Europe, and large population centers in Latin America, have up to 4 times higher levels of pollutants.

Gases are typically measured in ppb (parts per billion). In a dry standard atmosphere (760 Torr pressure) 1 ppb equals 2.46×10^{10} of molecules per cm^3 [1]. Thus 24.6 billion molecules must be multiplied by the total count in the above table (about 200). These numbers represent an arithmetic mean. Spikes in these numbers can cause these counts to climb by as much as a factor of 20. At standard atmospheric pressure (760 Torr) the kinetic theory of gases shows more than 10^{23} molecules will collide with each square centimeter of surface per second [13].

There are thousands of organic compounds in the air. Currently 151 are being measured. Of these over 40 are polarized organic acids [1]. Individually the amounts are not large but when the total amount is combined the result is up to 500 to 1000 µg per m^3 in a typical urban environment. These organic compounds are a major source of halogens: fluorine, chlorine, iodine, and bromine. In coastal areas airborne sea salt will also contribute to the presence of halogens. These halogens are released as a result of reactions with various acids, gases, moisture, and UV light.

When combined with moisture, carbon, silicon, other organics, and trace metals, a variety of contaminating materials can be created. In addition to sulfuric acid, nitric acid, hydrochloric acid, organic acids, various ammonium and sodium compounds are formed. These compounds are deposited on surfaces over extended periods of time gradually increasing the ability of materials to capture and retain moisture/water vapor more efficiently.

Moisture is a very effective solvent and catalyst. Relative humidity (RH) is the most common method of measuring the amount of moisture in the air. The number of molecules of water vapor in air with the following parameters, 25 degrees C, 760 Torr, and 50% RH, is approximately 4×10^{17} per cm^3. There is always a large amount of water vapor available for chemical reactions [1].

Many types of airborne particles and gases react with moisture and/or ultra-violet light by combining to form new chemical compounds and by being suspended or dissolved in airborne water droplets. A good example of this is the summertime haze seen in the northeastern United States. Up to 50% of this haze is a combination of sulfuric and nitric acids. These acids are a result of SOx and NOx reacting with various other gases, moisture, and ultra-violet light [2]. A "mature" acidic atmosphere contains 63% sulfuric acid, 31% nitric acid, and 6% hydrochloric acid (inorganic acid ratio) on the East Coast of the USA. On the West Coast of the USA the inorganic ratio can reverse with nitric acid almost doubling the amount of sulfuric acid. Organic acids in the air can also account for as much as 90% of the total suspended acids depending on the time of year and location.

During various times of the year when lower temperatures, moisture, and ultra-violet light levels limit the formation of atmospheric acids, sulfates and nitrates plus other suspended particles and gases collect on surfaces as dry deposition. Dry deposition removes from 60% to over 90% of all forms of contaminates in the atmosphere [1]. NOAA Air Turbulence and Deposition Division recorded the following amounts of gas dry deposition in the Ohio River Valley for one year in 1997: SO_2 97 $\mu g/cm^2$, HNO_3 46 $\mu g/cm^2$, and SO_3 6.7 $\mu g/cm^2$.

Moisture can be in three forms; water/rain, condensation, or high humidity/vapor. Generally, the only form of liquid the circuitry will encounter is condensation or spills. Any contaminants dissolved or suspended in water will collect near the surface of the drop. As the drop evaporates these contaminants will be drawn to the water/substrate boundary [3]. This creates a concentrated ring of pollutants (water drop rings). Because these rings are hygroscopic they will become the focus of moisture absorption, coating damage, and circuitry contamination.

Low pH factors, between 2 and 3, can occur when high humidity or condensation creates an aqueous layer that will dissolve/dilute contamination deposited on the surfaces. When the relative humidity reaches 60% a layer of moisture 2 to 4 molecules thick may form, and chemical reactions can begin. At 80% relative humidity the aqueous layer is 5 to 20 molecules thick and will begin to act as bulk water. Ions will begin to freely flow on the surface. Carbon can also be present on the surface. When combined with moisture and acids a corrosive cell can be established with the carbon material acting as the cathode. When voltage is applied to the circuitry the corrosion-degradation process will increase [4].

ISO 9223 defines "Time of Wetness" as a relative humidity above 80% at any temperature above 0 degrees C. In an effort to further quantify "Time of Wetness" Table 9.2 provides examples of hours per year in various climates.

Table 9.2 Time of wetness examples (from the Swedish Institute of Corrosion SB102)

Hours/Year	% of Year	Example of Occurrence
<10	<0.1	Indoor air with climate control
10-250	0.1-3	Indoor air in normal rooms for living or working conditions
250-1000	3-10	Indoor air storage rooms
1000-2500	10-30	Indoor air in some production rooms. Outdoor air in cold zone, dry zone, and parts of temperate zone.
2500-5500	30-60	Outdoor air in parts of temperate zone and parts of warm zone. Indoor air in animal houses
>5500	>60	Outdoor air in tropical zone. Indoor air in greenhouse

Under certain atmospheric conditions, an adsorbed water film is formed on the coating surface. The diffusing compound is presumed to be adsorbed on the coating film surface, dissolved in it, defused into the film, and released or desorbed on the other side of the film. For example, sulfur dioxide contained in the atmosphere can dissolve in this water film and can penetrate through the coating as far as the metal surface [4].

Any ions that reach the surface of the substrate can react with moisture and the metal circuitry to cause electrochemical migration when voltage is applied to the circuit. This can result in leakage currents, dendritic growth, conductive anodic filament (CAF) growth, cross talk, accelerated corrosion, plus possible shorts, arcing, and a potential ignition source [7].

Another factor to consider when dealing with moisture is CRH (Critical Relative Humidity). CRH is defined as the temperature and humidity level at which a material will begin to absorb water vapor from the air. When CRH is reached the particle will begin to grow. This process is known as deliquescence and the particles are sometimes referred to as aerosols. Aerosols are one of the main sources of visible haze.

**Table 9.3 Critical Relative Humidity (CRH) for several
inorganic compounds [6], [7]**

Compound	Temperature (°C)	Relative Humidity (%)
LiCl·H$_2$O	20	15
KF	100	22.9
KBr	20	84
KBr	100	69.2
KCl	0	88.6
KCl	80	78.9
CaCl$_2$·6H$_2$O	24.5	31
CaCl$_2$·6H$_2$O	5	39.8
(NH$_4$)HSO$_4$	24	40
NH$_4$·NO$_3$	24	62
(NH$_4$)$_2$SO$_4$	24	81
(NH$_4$)$_3$H(SO$_4$)$_2$	25	69
NaBr	100	22.9
NaCl	20	75
NaNO$_3$	25	74.3
NaCl·NaNO$_3$	25	68
NaCl·KCl	25	72.7
Na$_2$SO$_4$	25	84.2
NaF	100	96.6
H$_2$SO$_4$	25	5

Table 9.3 shows that the CRH for certain compounds will vary with temperature. When temperatures increase, molecules become more energetic, chemical reactions occur more rapidly, the permeability of the coatings increases, and osmotic pressures in the coatings increase. Over extended periods of field conditions these factors can combine to degrade reliability. Faults are usually intermittent and virtually impossible to isolate without very expensive laboratory testing.

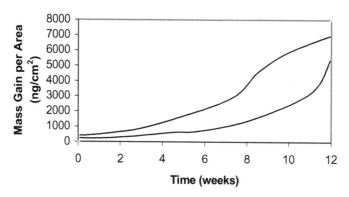

**Figure 9.2 Maximum and minimum weight gain on a gold coupon due to
particles and moisture [119]**

The relationship between particle deposition and moisture adsorption was assessed by M. Forslund and C. Leygraf, using gold coupons and measuring the mass gain in an outdoor sheltered environment (see Figure 9.2). The gold coupon was protected from direct exposure by a porous polytetrafluoroethylene filter, having a pour size of 10 microns. During the 12 week exposure period the amount of moisture measured on the surface of the gold increased from about 150 ng/cm^2 at >90% RH to 4000 ng/cm^2 while the amount of particulates increased from 0 ng/cm^2 to over 3000 ng/cm^2 (for reference 30 ng/cm^2 of moisture equals one monolayer of water) [12].

Particles are generally classified into four groups: organics, marine, soot carbon, and inorganics based on measurements of the core materials [1]. At least 30 to 40% of the core material must be in one of these classes to qualify. Some cores are too well mixed to determine their category. Others change their chemistry during the measuring process. A typical inorganic core can consist of ammonium sulfate plus sodium sulfate [1].

Table 9.4 shows the average fine-particle composition in urban and non-urban areas for the continental United States. These are averages and as such are subject to deviations depending on location, time of year, and local weather conditions.

Table 9.4 Average fine particle composition (0.01 to 2 micron diameter) [119]

Urban		Non-urban	
Material	**Percentage**	**Material**	**Percentage**
C-organic	31	C-organic	24
SO$_4$	28	SO$_4$	37
N/D (not determined)	18	N/D (not determined)	19
C-elemental	9	C-elemental	5
NH$_4$	8	NH$_4$	11
NO$_3$	6	NO$_3$	4

According to a Bell Laboratory report [3] the total dry deposition rate for particles in a standard outdoor urban environment is up to 2×10^9 per ft^2 per second (for .01 to 2 micron sizes). Indoor deposition depends on numerous factors including quality of the filtering systems, leakage of outdoor air, and indoor generated contaminates. The deposition rates can vary from 20% to as high as 80% of the outdoor rates.

Figure 9.3 shows the relationship between leakage current and relative humidity. These tests were performed on circuitry coupons having a 0.050-inch spacing (1.27mm) between component leads. The Kuwait samples were exposed to the soot and smoke from the oil fires during the Gulf War. This smoke had high levels of graphites and salt brine. The four samples in the lower right corner of the chart are "clean" coupons that were not exposed to airborne particles.

As the spacing/pitch between leads and traces is reduced opportunities for leakage current will rise. Conversely, the amount of voltage required to create the same level of leakage current will lower. The spacing/pitch between leads can be as low as 0.010 inch (0.25mm) and between traces as low as 0.005 inch (0.13mm).

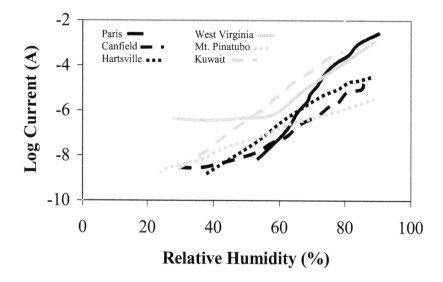

Figure 9.3 Leakage current of sample circuitry as a function of RH when exposed to indoor air in six locations: Paris, Canfield, West Virginia, Huntsville, Mt. Pinatubo, Kuwait [119]

The increased use of internal forced air-cooling, brought on by increasing power densities in electronics, has caused up to a 100-fold increase in the deposition rate of fine particles. In recent years, ionic contamination from fine particles has been a common cause of field failures. In fact, Airbus Industries issued a technical follow up bulletin (number 212600011) concerning problems with dust causing failures of computer systems that receive forced-air cooling. Quoting from the bulletin issued in June, 1993 [115]:

"Several failure cases of computers, mainly SFCC (Slat Flap Control Computer), installed in the Avionics bay compartment have been reported from in-service aircraft. Examination of subject computers revealed failure to be caused by intermittent internal computer board short circuits due to the presence of conductive dust inside the computer. Preliminary dust sample laboratory analysis showed presence of particles of sodium, iron, silicon and also small amounts of electrically conductive elements such as aluminum and titanium and small traces of calcium nickel, chromium and magnesium. The particles size ranged from 10 microns to over 200 microns with the largest particles being aggregates of smaller particles. Additional analyses of accumulated dust are being launched to determine/survey extent of phenomenon and also to identify origin/possible cause of subject contamination."

Although the dust may be conductive there are other factors to consider. Honeywell issued a service bulletin in February of 1997 that illustrates the complicated chemistry involving dust [116].

"SB No. M21-3341-037-00 - Modification: To protect against IRU (Inertial Reference Unit) failure when electrically conductive contamination such as chemical salts are present in the IRU and the IRU is operating in an environment which includes water condensing as a result of conditions reacting with the equipment cooling air."

Table 9.5 Nature and origin of common airborne contaminants [149]

Contaminant	Origin
Salts, acids, fats, dander	Operator sneezing, coughing, etc.
Salts and salt compounds	Outside air into the plant
Organic solvent vapors	Cleaning solvents, vehicles for plastics, and flux
Plasticizers	Heated plastics, heat shrink operations, vacuum forming
Pyrolysis products from plastics	Thermal wire stripping, grinding of plastics, thermal hot spots, accidental overheating
Isocyanates, amines, and other curatives	Mixing, applying, and curing of plastic processing compounds, ovens
Mold release agents	Plastic processing, especially when release agents are applied by spraying
Antistatic agents	Antistatic treatment by spraying
Oil	Solder shield fluid, vacuum pumps, compressed air
Rosin/resin	Fluxing and soldering
Metal fines	Grinding/machining operations
Common particulate contamination	Abrasive blasting, grinding, drilling

Designing for the end use environment requires an understanding of the widely divergent possibilities. The contractual requirements (reliability and duration) plus the cost of the product will determine the overall level of effort given to address these issues.

9.1.5 Tools

Whenever printed circuits are handled, all tools, handling fixtures, and other similar articles should be kept clean. It is useless to take precautions against circuit contamination, only to contaminate them with the tools that are used in assembly, such as lead formers, clippers, component placement machines, or reflow ovens [43].

9.1.6 Fault Markers

It is common during inspection processes to mark printed circuits with self-adhesive arrows, felt-tip pens, or graphite pencils wherever a fault has been noticed. Self-adhesive products should leave no adhesive when removed. Not only is the adhesive a contaminant itself, but also dust settling on the adhesive can make the situation even worse. If felt-tip pens are used, tests should be conducted to ensure that all the resulting residues, including the dyes, can be easily removed. Some felt-tip pens have highly penetrating acidic inks [43].

9.2 Storage

When printed circuits and their components are stored, they are frequently in open bins. Putting printed circuits into a hermetic container on a hot humid day is just as bad if the temperature drops. Critical printed circuits should thus be stored in closed containers with a small silica gel bag (of course, the silica gel bags should be desiccated prior to reuse). If these precautions are taken, it is not necessary to have a primary wrapping around the individual printed circuit, although this could alleviate some of the difficulties caused when the circuit comes into intimate contact with an inadequate or unsuitable wrapping material [42].

Plastic containers or bags made of polymers with high plasticizer content may introduce contamination. This can occur when containers are exposed to heat (such as from direct sunlight) resulting in plasticizer bleed-out, evaporation, and recondensation on the contained parts. This is similar to the yellowish deposit formed on the windows of a car with vinyl upholstery after long periods on a hot day with the windows closed. The plasticizer from the vinyl seats evaporates and condenses on the windows. The same phenomenon has been known to occur with components and circuit boards wrapped in or contained by plastic bags or boxes [42]. Corrugated cardboard containers are commonly used for shipping and storage. These boxes must not contain sulfur or sulfate byproducts.

Precautions should be taken concerning the material in the storage area. Unsuitable floor coverings include carpet, linoleum, vinyl tiles or sheets, bare cement or concrete, or any floor covering that cannot be kept clean. One of the best floor coverings is oak parquet that has been impregnated with a synthetic resin prior to being laid. This can be readily washed for cleaning. Synthetic materials should generally be avoided for shelving and storage, unless they are specifically designed for this application and guaranteed by the manufacturer to be free of contamination-producing effects [42].

9.3 Components

Component contamination affects solderability and can possibly spread onto other areas of the assembly via handling and cleaning operations. Typically, contamination on component leads consists of fingerprints and surface oxidation products that are capable of reducing solderability. Time and storage conditions greatly influence the formation of these oxidation products [46].

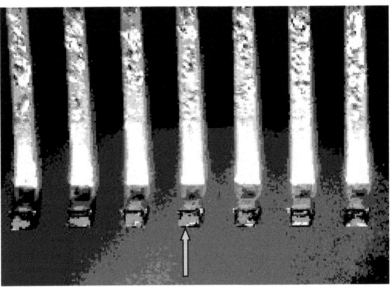

Figure 9.4 Contamination on QFP leads (possibly lubricant from lead trimmer). With permission from CALCE Electronic Products and Systems Center.

In general components do not contaminate. However, mold release agents encountered on component bodies and plastic cases can interfere with operations such as component bonding and conformal coating. Inks used for identification may not be compatible with the processing compounds used and may flake off and become redeposited in electrical contact areas of the finished product. Bromine, which is added to the plastic encapsulant to render it flame retardant, can outgas bromine compounds during high temperature operations such as during soldering [46]. Inadequate cleaning after soldering may leave flux residue. Aged rosin flux will be difficult to remove by the common assembly cleaning process.

Most semiconductor devices are encapsulated in a molding compound to provide environmental and physical protection for the device, aid in assembly pick and place, marking, and in some cases, heat transfer [141]. However, encapsulants do not act as complete barriers to the ingress or diffusion of water, oxygen, ions and contaminants. [117] Unlike ceramic and metal packages, the encapsulants are not considered hermetic.

In the past decade, a number of techniques have been used to study moisture absorption in encapsulants including: gravimetric techniques, dimensional change, fluorescence measurements, neutron scattering, permittivity and capacitance measurements, IR spectroscopy, and Proton NMR spectroscopy [117]. In addition, there are many models for the diffusion of moisture through microcircuit encapsulants. For example, Haleh et al. describe typical encapsulants used in microcircuit packaging and moisture diffusion in these encapsulants. Several books and review papers have been published on diffusion in polymers [136-138] and the effects of moisture on electronic systems [139-142].

The potential for corrosion of electronic materials is determined by the nature of the local environment. Factors influencing corrosion include: humidity, temperature, electrical potential, materials, and contaminants. Because corrosion of the device metals requires the presence of water, a reducible species and an electrolytic path for ion transport, the elimination of any of these components will prevent the corrosion process.

Moisture diffusion in encapsulants can be a reliability concern, if the packaged device is exposed to contaminants from fabrication process, the packaging materials or from the external environment. However, the device fabrication process has become quite clean. Today it is rare that corrosion arises due to contaminants inherent in the fabrication process. Thus the issue addressed in this paper is primarily the transport (diffusion) of ions through the encapsulant from an external source.

Encapsulants used to package microelectronic devices are formulations of eight or more different materials. The base polymers used in these formulations are typically epoxy cresol novolacs or epoxy biphenyls. These polymers are hydrophilic and inherently absorb moisture. However, it is not the moisture alone, which leads to corrosion in integrated circuits but the presence of ions in the absorbed moisture [117-119]. The presence of ions is usually the most important factor in the corrosion of electronic materials and devices when moisture is present [118].

Several studies have been published on moisture absorption in epoxy molding compounds used for microelectronic encapsulation [120-128]. The effect of ions in encapsulants on corrosion of bond pads and circuit metallization has also been reported [129-135]. Chloride ions from resin synthesis and bromide and antimony ions from added flame-retardants have been identified as the primary corrosive ions in the encapsulants. The primary site of corrosion in encapsulated devices is the bond pad, which is not protected by a passivation coating and is thus most susceptible to attack by corrosive materials in the molding compound. However, the ions must diffuse to the bond pad area in order to react with the bond pad metallization. Thus the ability of the ions to diffuse in the encapsulant is expected to play a role in the rate of corrosion and the reliability of encapsulated devices. There is evidence in the literature that the rate of corrosion-induced failures in encapsulated devices varies with the molding compound formulation. However, ion diffusion rates in microcircuit encapsulants have not been reported.

9.4 Assembly

The assembly process is a potential source of contamination. This section covers some of the major steps of printed wiring assembly and how each step can contribute to contamination. This is not meant to be an all-inclusive list of potential assembly process contamination sources, but does cover a basic list of process steps and the associated contamination risks.

9.4.1 Temporary Masking

When the printed circuit boards reach the assembly line, one of the first operations is to apply a temporary mask to contact areas and pads without leads to keep them from being soldered during the mass soldering process. As a general rule, four different types of temporary masks are used. The first is a self-adhesive tape; the same concerns apply here as to fault-indicating arrows. The adhesive comes off without leaving any residue whatsoever, even after the soldering process.

The second type of temporary mask, used mostly for edge connectors, consists of a titanium clip that prevents the gold-plated contacts from touching the molten solder. If the clips are cleaned after each use, no problems should arise from their use. However, care must be taken when applying or removing them not to scratch the contacts.

The third type of temporary solder mask is a latex compound, generally of either an ammoniacal or a silicone base. Silicone base compounds should be avoided, since they apply a class G contamination to the circuit. Ammoniacal-based latex compounds, which have a distinctive smell when wet, must be applied in such a way that, after soldering, the temporary mask can be easily peeled. If it tends to tear or stick to the circuit during removal, it should be replaced with another type. However, temporary solder masks of the ammoniacal latex type can cause ionic contamination that is not removable by solvent cleaners of any type. With this class C or D contamination, an adequate water rinse, preferably with a suitable detergent, should ensure a thorough removal of the alkaline contamination.

The fourth type of temporary solder mask is the water-soluble solder stop. These are usually solutions of natural gums combined with some kind of thickening agent. In theory, after soldering, the residues are completely water-soluble. Water-soluble solder stop is usually compatible with the use of water-soluble fluxes followed by a water wash or, alternatively, with the use of resin fluxes, where flux removal is ensured by a saponification process. However, it requires an extremely thorough hot water wash to remove the residues from these temporary solder masks. For this reason they are generally highly colored, so that any residues that may remain after washing are readily visible and indicate that further washing is necessary. If correctly removed, then these products are almost without any risk.

9.4.2 Component Mounting

During component mounting, the possible contaminants depend on the method employed. If the heads of the machine and the component forming guides are clean, fully automatic component placement using numerically controlled machines offers the least risk of contamination. However, such machines require axial components to be taped into rolls. If these components are contaminated it is very difficult to provide a suitable decontamination process before mounting. This means that any decontamination must be done after the dirty components are mounted onto the board, soldered, and cleaned. Provided that the contamination does not seriously affect the solderability of the component, this can be done effectively.

With semi-automatic and manual mounting, it is usual to form the leads of the components as a separate operation before assembly. If the components are affected by any kind of contamination other than class G, they can be bulk-cleaned prior to mounting onto the boards. If the soldering flux is not removed later, it is essential that both the components and the printed circuits be decontaminated and tested before assembly.

The use of permanent and temporary adhesives to install surface mount components prior to soldering is a growing practice in the industry. Compatibility with other materials as well as the processes used must be well known before full-scale production begins. Underfills and "Glob-Top" materials must also be tested.

9.4.3 Fluxing

The flux itself, being a deliberate contaminant, will not be discussed at this point, but a number of contaminants can be inadvertently introduced during the fluxing process. With normal good housekeeping, the contaminants accidentally introduced during fluxing are seldom serious [54].

The most common contaminant in the flux is introduced by an accumulation of foreign matter from the printed circuit boards themselves. This can be either particulate matter or any contaminant that is soluble in the solvents used in the flux. The only remedy for this kind of contamination is to change the flux regularly and to clean the fluxer at the same time [54].

The second type of contamination introduced during fluxing comes from compressed air used either for foaming the flux or for blowing away excess flux. This contaminant frequently involves lubricating oil from the air compressor and water. The latter is very insidious, particularly if resin or rosin-type fluxes are used in an ethyl alcohol base. The ethyl alcohol itself is hygroscopic and will ensure that all the water introduced by the compressed air will remain in the flux. Water will hydrolyze pure wood rosin, resulting in the application of partially hydrolyzed rosin onto the printed circuit board [54]. Among other things, this renders the rosin practically impossible to remove.

A third type of contaminant introduced into the flux comes from the chariots or finger conveyors that transport the boards. This can be particularly severe if oil is used on the solder surface, especially with oil intermix systems on wave soldering machines. With finger conveyors, the fingers are being continually fluxed, passed through the wave (with consequent thermal degradation of the flux and other products), returned back through the flux (in some machines the returning fingers may be lifted clear of the flux), and then back to the flux again. The flux will become contaminated unless these fingers are cleaned regularly. Obviously, with chariot conveyors the degree of contamination is much less since chariots pass only when circuits are loaded in them for soldering and they do not return backwards through the flux. Some form of cleaning of the circuit holders between two passages in the machine is useful.

Flux residues that are not removed during cleaning operations can cause contamination. These residues can degrade bonding [86]. Organic contamination such as oils, fats, and greases may also be dispersed throughout the flux residue and may cause a similar problem when aqueous defluxing is applied after soldering with water-soluble flux [4]. These substances are not soluble in plain water and can only be removed when the aqueous defluxing agent also provides saponification and dispersion effects, or when additional organic solvent cleaning is applied. Table 9.6 shows contaminants that typically can become dispersed in the flux residue [4].

Table 9.6 Nature and origin of contaminants contained in flux residue [149]

Nature of Contaminant	Solubility	Origin	
		Surface	Flux
Cupric halides	Water	Copper	Halide activators, halide flux
Cuprous halides	Ammoniated solutions	Copper	Halide activators, halide flux
Copper abietate	Insoluble	Copper	Rosin
Nickel halides	Water	Nickel or tin-nickel	Halide activators, halide flux
Stannous halides	Acidic aqueous agents	Tin-lead or tin-nickel	Halide activators, halide flux
Stannic halides	Water	Tin-lead or tin-nickel	Halide activators, halide flux
Lead halides	Water	Tin-lead	Halide activators, halide flux
Polyglycol	Water	Solder wave	Solder shield fluid
Oils, fats, greases	Organic solvents	Any	Solder shield fluid, fingerprints, condensed oil fume

Soldering flux is defined in the IPC flux specification ANSI/IPC-SF-818 as "*a chemically and physically active formula, which promotes wetting of a metal surface by molten solder, by removing the oxide or other surface films from the base metals and the solder. The flux also protects the surfaces from reoxidation during soldering and alters the surface tension of the molten solder and the base metal.*" The vast majority of soldering is done with the aid of fluxes. Any necessary post-soldering cleaning process will depend on the flux used in the soldering process.

Fluxes are composed of a number of materials, including solvent, vehicle, activator, surfactant, and antioxidant. The solvent is the liquid carrier for the flux ingredients, allowing even distribution of the flux material on the board. During preheating, the solvent evaporates, so little is present when the board contacts the solder wave. For most fluxes, the solvent is isopropanol. The vehicle is a thermally stable material that acts as a high-temperature solvent during wave soldering. Sometimes it is also a weak activator, such as rosin. Depending on its properties, it can inhibit or induce corrosion.

The activator consists of one or more ingredients in a flux that creates a wettable surface for solder by removing oxides and other contaminants when in contact with the metal on the board. These ingredients may or may not be corrosive at room temperature, but certainly must be active or corrosive at elevated temperatures to perform their job properly. Activators are typically organic solids. The surfactant reduces the surface tension at the metal/solder interface to further promote solder wetting, especially when the printed circuit board exits the solder wave. It can also serve as a foaming agent, sometimes a necessary additive so a flux can be applied with a foam fluxer. The antioxidant prevents reoxidation of the metal surfaces after the activator has prepared them for soldering. Often, another material, such as the vehicle, might also serve as an antioxidant.

Fluxes can be broken down into four categories: rosin, water soluble, synthetic activated, and low solids. Before looking at each group individually, it is worthwhile to compare the makeup of these fluxes in terms of generic constituents and proportions, looking only at the three major components- solvent, vehicle, and activator. Rosin fluxes typically have a much lower activator-to-vehicle ratio than other fluxes. Water-soluble and synthetic-activated fluxes may have about the same proportion of activator relative to the amount of vehicle. The low-solid fluxes have much less solid matter, and the amount of activator is typically the same or more than the amount of vehicle.

According to J-STD-004, "Requirements for Soldering Fluxes," all solder fluxes can be broken down into three basic types:

∞ Type L – all nonactivated rosin types, most mildly activated rosin (RMA), and some activated rosin (RA) fluxes- i.e., all rosin fluxes with low to no flux activity

∞ Type M – some RMA, most RA, some water-soluble and synthetic-resin activated fluxes- i.e., all fluxes with moderate activity

∞ Type H – all superactivated rosin fluxes, most water-soluble and synthetic-resin activated fluxes- i.e., fluxes with very high activity

These classifications are assigned through the results of the copper mirror test, the silver chromate and halide content test, and the British corrosion test. These tests determine the corrosive potential of the flux. An additional test is flux characterization with surface insulation resistance (SIR) measurements. Because of further groupings for SIR testing, fluxes categorized as L, M, or H are followed by the numeral I, II, or III. Fluxes identified I or II have passed the SIR requirements for assembly classes I and II (general industrial requirements); III indicates passing the SIR requirements for assembly class III (military and other high-reliability circuits). The letters C, N, or CN are assigned depending on whether the flux meets the SIR after being cleaned (C), not cleaned (N), or both (CN). The letter O affixed to the flux type shows that either the flux has not been tested or that it cannot meet the minimum SIR requirements. The requirements for each test are outlined in Table 9.7.

Table 9.7 Recommended classification of fluxes according to J-STD-004

Type	Test	Requirement
L	Copper mirror	No evidence of copper removal or partial removal indicated by white background showing through in part of the fixed area*
	Silver chromate or halide	Pass silver chromate test** or "0.5% halides
	Corrosion	There should be no evidence of corrosion. If a blue/green line occurs at the flux metal interface, the area of corrosion shall be tested per silver chromate test and must pass those requirements
M	Copper mirror	Partial or complete removal of the copper mirror in the entire area where the flux is located
	Halide	"2.0 % halides
	Corrosion	Minor corrosion is acceptable, provided the flux or flux residue can pass the copper mirror or halide test for this category. Evidence of major corrosion places the sample or flux into the H category
H	Copper mirror	Complete removal of the copper mirror in the entire area where flux is located
	Halide	>2.0% halides
	Corrosion	Evidence of major corrosion

*Discoloration of the copper or reduction in the copper film thickness is not sufficient to remove the flux from this category.
**Failure to pass the silver chromate test requires that the flux pass the halide test to be considered for category L.

Copper Mirror Test: The copper mirror test referenced in a number of specifications is a measure of a flux's corrosivity to copper. Corrosivity is an indication of the activity level of a flux. Both the speed and the relative ease with which this test can be performed make its use quite common. However, it has several drawbacks.

The copper mirror test involves placing two drops of the flux in question on a copper mirror, a glass microscope slide onto which 300 to 500 Angstroms of copper is vapor-deposited. Typically, a flux made of pure water white gum rosin and isopropanol acts as a control a drop of it is also placed on the mirror. The mirror is placed in a controlled environment of 23±2℃ and 50±5% relative humidity for 24 hours; then rinsed in isopropanol and examined for any spots or areas where the copper has been removed by the flux [47].

Results of the copper mirror test indicate the potential corrosivity of the raw or unheated flux; the test unfortunately does not always indicate the corrosivity of a partially or fully heated flux. The results can be difficult to interpret, since the test is somewhat subjective and there is no middle ground between pass and fail. Because of these drawbacks, critics have questioned the relevance of this test altogether [47]. However, the speed, relative ease, and evidence that its results correlate well with those of other lengthier and more involved tests, justify the worth of this test. It is especially useful for prescreening when testing a new formulation, or as a routine incoming materials check.

Silver Chromate and Halide Test: The silver chromate paper test is another fairly simple test. Its purpose is to detect significant amounts of chlorides and bromides in a flux. Any flux that leaves the silver chromate paper unchanged has either no Chlorine⁻ or Bromine⁻, or at least no detectable quantities [47].

The test involves placing a drop of flux on silver chromate paper. After 15 seconds, the paper is rinsed or immersed in isopropanol, the allowed to dry for 10 minutes. A color change means that halides are present in detectable quantities [47].

Drawbacks of this test include possible subjectivity in interpretation of results, lack of quantitative data, and interference from other non-halide materials. Some chemicals can cause color change, as can fluxes that are unusually acidic. Thus, if halides are not suspected and a color change appears, other tests should be performed to verify the results [47]. Another problem with this test is that the results are pass/fail only, and not quantitative. A proposal in a telecommunications specification, for example, has called for a more quantitative assessment of the amount of chlorides and bromides in a flux, based on a more specific color change in the silver chromate paper. Yet another problem is that the silver chromate paper only detects chlorides and bromides, but not fluorides. Even with these shortcomings, this test is still beneficial for prescreening or incoming material inspection for fluxes that should be non-corrosive.

British Corrosion Test: The British corrosion test involves reflowing solder in the dimple of a copper panel in the presence of flux solids, then aging the panel at 40°C and 93% RH for a number of days, depending on the specification and type of flux being tested. A flux passes if no blue-green corrosion can be observed after aging [47].

The test has both benefits and drawbacks. On the plus side, the results are independent of a flux's solids content, since this is held constant. Because a concentrated flux material is used, the test constitutes a worst-case condition [47].

However, the test lacks an electrical bias, which is necessary to initiate some corrosion mechanisms. It creates an accentuated copper/solder/flux interface. Lastly, the interpretation of results is difficult, subjective, and only quantitative. Evaluating the extent of green or blue-green coloration is difficult. As with the copper mirror test, there is no middle ground between pass and fail. The difference between this test and the SIR test lies in their distinct purposes; the corrosion test evaluates the potential for an open failure, while the SIR test determines the potential for an electrical short [47].

9.4.3.1 Rosin

The vehicle in rosin fluxes is rosin or colophony, a naturally derived material from pine trees. In most places in the world, the word *rosin* usually refers to the natural substance, whereas the word *resin* refers to a synthetically made material. However, sometimes the words are used interchangeably. The rosin normally used in fluxes is white water rosin, a high-grade material similar to compounds used in varnishes and lacquers.

Rosin material has several unique characteristics. Besides acting as a flux vehicle, it is also a mild activator at soldering temperatures. Rosin consists of a number of isomers, mostly abietic acid; these abietic acids are fairly large and include one carboxyl group ($C_{19}H_{29}COOH$). Another unique property of rosin is that it is an extremely good insulator at room temperature; in fact, its bulk resistivity is at least an order of magnitude higher than that of epoxy-glass. Because rosin hardens on cooling to room temperature, it serves as an excellent encapsulant, prohibiting the movement of other flux residue ingredients, such as those from activators. However, at soldering temperatures, rosin is mildly active; so when dissolved in an appropriate solvent, such as isopropanol, it can perform all the necessary functions of a flux. Usually the activity of rosin by itself is not sufficient for production soldering most components and printed circuit boards [49].

It is important to note that rosin will undergo a change around 75℃ to 85℃, its softening point, which can impact accelerated aging or surface insulation resistance tests performed at or above this temperature range. By softening, the rosin loses its encapsulating properties, allowing usually tightly bound flux ingredients to be more mobile. Sometimes rosin flux residue is erroneously categorized as harmful because it has been tested above its softening point. If a printed circuit board will never see operating conditions this high, then a failure mode observed during high temperature aging conditions is irrelevant [49].

Because of the insulating properties of flux residue, it is not always cleaned off. In fact, many companies leave rosin flux residue unremoved for years without problems. Nevertheless, there are a number of reasons for removing it. Dust and dirt attaches to the assembly easily because rosin residue is sticky and tacky, therefore, it may have to be cleaned. Another reason for cleaning is to avoid problems during bed-of-nails or in-circuit testing. Rosin flux residue can cause false opens when the test pins are unable to push through the insulating residue covering a test pad. Sometimes it is necessary to remove rosin residue because of its effect on the subsequent addition of a conformal coating. Unfortunately, completely removing rosin residue is often quite difficult [59].

9.4.3.2 Water Soluble

Water-soluble fluxes (WSFs) are typically very active and aggressive, and the residue can be corrosive if not removed properly. However, they offer a number of advantages over less active and more difficult to remove fluxes. In fact, with their use the process window for soldering is relatively large, and extremely low solder defect levels are attainable [48].

Often WSFs are referred to as organic acid (OA) fluxes. Although there are organic acids in many WSFs, other fluxes, including rosin, SA, and low solids, also include organic acids. Thus, naming water-soluble fluxes OA is misleading. The vehicle for WSFs is typically a polyethyleneoxide or polypropyleneoxide, commonly referred to as polyglycols. These materials are nonionic and do not encapsulate activator residues like rosin [48].

Usually the activity level is quite high in WSFs, meaning that fairly strong, or relatively large quantities of organic acids and halide bearing compounds are present. Some WSFs are halide-free and promoted as less corrosive, yet their postsolder residue still needs to be removed because of its organic acid [48].

WSFs, like other aggressive fluxes, must be removed in a spray impingement or other energy-enhanced cleaning system. The cleaning medium should either be aqueous only or an aqueous detergent saponifier solution in an inline system.

9.4.3.3 Synthetic Activated

In the 1980s, Dupont developed a family of high activity synthetic activated (SA) fluxes, analogous to WSFs, with residues designed to be easily soluble in CFC-113/methanol azeotrop. SA fluxes are based on a mono- and diisooctyl phosphate (IOP) chemistry and include compatible activator and solvent materials.

Because of their high activity, these fluxes perform much like WSFs, and will not pass the copper mirror or silver chromate paper tests. Also, SA flux residues must be removed in a spray impingement or total immersion cleaning system. Usually the cleaning medium is CFC-based, but terpenes also do the job adequately. Thus, component compatibility with the cleaning medium must be addressed. Unfortunately, a water-compatible component may not be compatible with CFCs, and vice versa. If the cleaning process is efficient, the requirements for SIR and corrosion tests can be met with SA fluxes [48].

Like WSFs, SA fluxes typically yield fewer solder defects than rosin fluxes, and their residues are more easily removable. Nevertheless, even with these benefits the use of SA fluxes is declining due to environmental concerns with CFCs.

9.4.3.4 Low Solids

Unlike other fluxes, low solids fluxes (LSF) are formulated to leave minimal or no postsolder residue. This category includes fluxes with a much lower solids content them traditional fluxes- from 2 to 5 wt. % [47].

Eliminating postsolder cleaning also eliminates machine cost and maintenance, and reduces materials and operating costs. In addition, space savings and process simplification are also realized. The need for any postsolder assembly steps is eliminated when components incompatible with the cleaning process must be manually inserted and soldered, and the associated temporary solder mask used as a solder resist for plated through-holes is not needed [47].

Concerns over EPA regulations, international restrictions, component qualification, and waste disposal are precluded by the elimination of cleaning. Environmental concerns are becoming increasingly important in process decisions.

The use of LSFs improves solder yields by as much as an order of magnitude over to rosin fluxes, though these changes are also dependent on their product code and process parameters. Because of the lighter consistency and freer flow compared to rosin fluxes, low solids fluxes can more easily wet all surfaces of the wiring side of a board, especially important when surface-mount devices are wave soldered.

Ideally, a flux that does not need postsolder cleaning should be noncorrosive, leave a minimal amount of residue, have sufficient activity for soldering, and be compatible with the process equipment. Unfortunately, it is difficult to find an LSF that meets all of these characteristics; the flux application process needs to be modified to satisfy all of the requirements.

When low solids content is maintained and sufficient flux activity insured, the activator often outweighs the vehicle. Because of this unusual ratio, the vehicle may not encapsulate activator residues, as is the case with more active LSFs. On the other hand, some flux manufacturers have made LSFs that are simply diluted versions of traditional noncorrosive rosin fluxes, in which the amount of rosin far outweighs the amount of activator, but their activity level is not adequate for many soldering operations [46].

The solvent used in LSFs is typically isopropanol, as in other fluxes. The only difference is that there is significantly more solvent in an LSF than in a traditional flux. The vehicle in LSFs is a synthetic resin material, rosin, or modified rosin. Those that contain rosin are amber in color, rather than clear, and can leave residues that impair bed-of-nails testing. Most LSFs are halide-free and contain only organic acid type activators. Those with halides usually also include rosin. Preferably, an LSF will be halide- and rosin-free [46].

One difficulty when using low solids fluxes is determining the quantity of flux to be used; enough should be used to ensure sufficient fluxing activity, but not so much that it leaves a lot of residue. The effect of water absorption can be more pronounced, and the specific gravity more difficult to control, than in fluxes with a higher solids content. The "no clean" concept applies only if the circuitry and all the components are "clean" prior to the soldering operation. These drawbacks create a narrower process window, so the selection of application equipment and process parameters is important. If these are chosen wisely, LSFs can yield great benefits.

9.4.4 Soldering

After the flux is applied, usually a preheating section dries off the flux solvents. If blown air is used, the air intake to the fan should be situated where chemical contamination cannot be drawn into it. Particulate contamination, especially important with sticky rosin or resin fluxes, should be eliminated with good filtering [13].

The soldering operation itself, particularly if oil is used on the wave, can introduce many different types of contamination. These include the reaction products between the flux, the oil, the substrate, the conductors, the solder and the air, and any contamination on the board or components before the flux is applied. In addition, certain components may thermally decompose at the soldering temperature; this is sometimes visible as charring of the flux or as discoloration of the substrate. Other decomposition products may not be so easily identifiable, notably polymerized compounds, which look like a monomer or partially polymerized material [13]. These heavy polymers sometimes cause problems due to their relative insolubility in solvents and water.

During the soldering operation temperatures between 240 and 260 degrees C are common using HASL (Hot Air Solder Leveling) or fusing. These temperatures are well above the Tg of most substrates. In the case of FR-4 type substrates the epoxy resin in the laminate becomes very soft at these temperatures. Flux ingredients as well as chemical compounds created by the soldering process can become absorbed into the substrate. Depending on the type of flux used these contaminates and compounds can range from acids, bromides, halides, chlorides, polyglycols, to carbon compounds plus microscopic metal particles and salts. Once these are absorbed into the substrate, they can come out during later assembly processes and initiate metal migration. To further complicate the matter many substrates have flame retardant compounds blended into the rosin. These usually consist of bromides, which will precipitate to the surface when the assembly is heated during soldering. This combination of contaminate materials is virtually impossible to remove with a single cleaning solution or operation.

This problem is aggravated if reflow or rework is done on the assembly. Besides the obvious laminate absorption, which happens numerous times as different types of flux are used at higher temperatures. A great deal of caution is necessary if the original flux was the no-clean type. It has been observed that different types of no-cleaned fluxes made by the same company are capable of corrosive chemical reactions when they come into contact with each other. Due to the wide variety of fluxes available careful compatibility testing is required if different types of fluxes are used for rework or reflow.

Soldering can originate a variety of contaminants, in addition to those discussed under fluxing. Although soldering and fluxing are performed in one operation, the contaminants resulting from soldering are discussed separately to emphasize their origin. Contamination from soldering can be solder dross, excessive amounts of metallic impurities in the solder, or oils from solder shield agents. Most of them are encountered when solder wells are used. Contamination from solder impurities does not occur with hand soldering, because the solder wire has controlled amounts of impurities and the melted portion is used almost instantly [13].

Solder wells keep large amounts of solder molten for dip, wave, or drag soldering. The solder in these wells is subject to surface oxidation and subsequent dross formation. In addition, there is a cumulative pickup of metal from the soldered surfaces.

Contamination can also originate from different combinations of metals in the solder [86]. Some examples are:

Aluminum- causes frosty appearance in solder joints and reduces solder adhesion

Cadmium- produces "sluggish" joints and reduces wetting and spreading

Copper- Can decrease the fluidity of the solder and cause a gritty and "sluggish" appearance in the joints.

Iron- may cause grittiness and form intermetallic needle-shaped crystals that float on a solder bath

Gold- in small amounts it can produce dull and sluggish solder and in large amounts it can decrease strength in the solder joint

Solder itself may become a contaminant when deposited on electrical insulator areas, where it reduces the dielectric spacing or causes a short circuit by bridging conductors; or it may be deposited over noble metal finishes, where it may cause galvanic corrosion during environmental exposure. Inadvertent solder deposition on gold fingers may occur from inadequately masking the fingers. Liquid solder, when contacting gold surfaces, will extract the gold and contain it in the solder deposit; the contaminant becomes occluded and practically impossible to remove [8].

Solder itself may become a contaminant. When deposited on electrical insulator areas it can reduce the dielectric spacing or cause a short circuit by bridging connectors. When deposited over noble metal finishes it can cause galvanic corrosion during environmental exposure. Another problem is the incomplete vulcanization of the activators in the solder. This problem is much more serious during repair or rework. The solder may move away from the heat source, become embedded in the laminate, under or in components, and is very difficult to remove. The nonvulcanized activators are very ionic and corrosive. Controlling the heat profile during manufacturing and using thick forms of solder during rework and repair can help control these problems. Inadvertent solder deposition on gold fingers may occur from inadequately masking the fingers. Liquid solder, when contacting gold surfaces, will extract the gold and contain it in the solder deposit; the contaminate becomes occluded and practically impossible to remove [8].

Table 9.8 Solder Alloy 63Sn/37Pb: Solder bath maximum allowable contamination levels, %

Specification	Al	As	Bl	Cd	Cu	Fe	Ni	Zn
QQ-S-571	0.005	0.03	0.25	--	0.08	0.02	--	0.005
MIL-S-46844	0.005	0.03	0.25	--	0.08	0.02	--	0.005
WS-6536 (Pretinning)	0.008	0.03	0.25	0.01	0.50	0.02	0.025	0.008
WS-6536 (Mach. solder)	0.006	0.03	0.25	0.005	0.25	0.02	0.01	0.005
DOD STD 2000-1B (Pretinning)	0.008	0.03	0.25	0.01	0.75	0.02	0.025	0.008
DOD STD 2000-1B (Mach. solder)	0.006	0.03	0.25	0.005	0.30	0.02	0.01	0.005
NHB 5300.4	0.005	0.03	0.25	--	0.20	0.02	--	0.008

A more detailed diagram of contaminants from substrate fabrication is shown in Figure 9.5. For classification of the contaminants, please refer back to Table 1 (pg. x) for each by-product.

9.4.4.1 Blowholes

Blowholes are caused by the presence of some gas or vapor inside a plated through-hole. The heat from the soldering process causes a bubble of vapor to blow through the fillet of molten solder to form hidden bubbles or, in extreme cases, a crater [8]. The following are possible causes for the formation of blowholes:

∞   imperfect drilling, causing tearing, rather than cutting the glass fibers;
∞   a nonstoichiometric ratio between the resin and the hardener;
∞   water entrapment;
∞   entrapped hydrogen during electroplating;
∞   resin smear;
∞   co-deposition of organic substances with the metals;
∞   insoluble organic material on the surface of the fused or leveled tin-lead;
∞   insoluble organo-metallic salts on the surface of the metal;
∞   certain types of copper cleaning compound before tin-lead plating.

The metallization process does not normally introduce contamination to the substrate unless the substrate is inadequately prepared. In addition, if agitation during metallization was sufficient, then the hydrogen released during copper reduction will be trapped in the substrate holes at the interface with the deposited copper layer. The entrapped hydrogen causes out gassing and degradation of the plated hole [4].

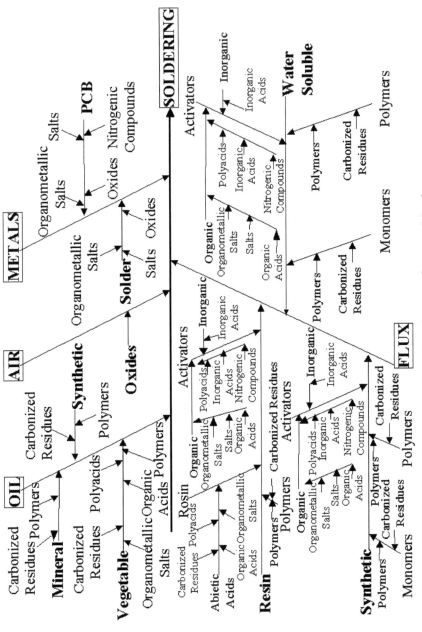

Figure 9.5 Detailed Diagram of contaminants from soldering process

9.4.5 Cleaning

Flux removal (discussed in later sections) should take off all the particles that remain on the board after soldering, but rarely does so completely. Neither solvents nor water alone can completely remove all the by-products of any flux. Flux removal is essentially a compromise between the removal of the greater part of the residues (and the most visible components) and other factors. Although the removal solvents and techniques may eliminate the bulk of the residues, however, they frequently introduce other new ones themselves [22]. In addition, if the circuit or components are contaminated with other substances before soldering (such as fingerprints), the reaction products will almost certainly not be removed by any one cleaning process. It is therefore unlikely that a circuit will be ultra-clean after conventional flux removal techniques [17].

The usual halogenated organic solvents can introduce contaminants onto an otherwise clean assembly via several mechanisms. The hydrolytic breakdown of halogenated solvents in the presence of flux residues produces acids [40]. Stabilizers are added to reduce the risk that solvent alcohols that have selectively absorbed water will attack light metals, such as magnesium, aluminum, beryllium, zinc, and their alloys. If this occurs, certain metallic compounds could also be present in the solvent [40].

Pure water is the only noncontaminating solvent, because it can never break down into other, more dangerous compounds. However, pure water does not exist in practice. Apart from contaminants dissolved in the water from the assemblies themselves, contaminants may be brought in by the water itself, using either a closed circuit or an open circuit. The closed circuit should be avoided, since the purification process will not eliminate some types of contaminants. These contaminants will accumulate in the water and cause more harm than good [41]. In particular, nonionic surfactants are notoriously difficult to remove and leave hygroscopic deposits on cleaned assemblies.

In an open circuit, the contaminants introduced by the water itself are clearly a function of water purity. Tap water will leave some of its dissolved and suspended solids on the parts being cleaned. This effect will be reduced if air-knife drying blows the water, along with its impurities, off the assembly.

If tap water is filtered and deionized, it should contain little dangerous contamination, particularly with air-knife drying. However, the deionizing columns are often a breeding ground for organic impurities, notably algae, which can leave vesicating deposits on the parts being cleaned. Water coming out of such columns should be considered suspect if the column has been in use for approximately six months, regardless of the throughput. If the water is passed through transparent pipework, light may cause a proliferation of algae spores [41]. It is a good idea to pass purified water through a viewing window that is kept illuminated night and day to show organic growths that are likely to become dangerous. Where this is a real problem, the water may be treated with a copper sulfate algicide before deionization, at the cost of reducing the useful capacity of the columns.

Other contaminants in water are produced by free chlorine. Most tap water is chlorinated to kill off pathogenic contaminants (bacteria and viruses). Chlorine is hydrolyzed by water to give chlorine hydrate, but this has a slight tendency to break down to give chloride and hypochlorite ions, both of which are active [41]. The ionogenic-free chlorine, being nonionic itself, can pass unhindered through any deionizing columns, to decompose afterwards.

Another contaminant that is a possible nuisance is atmospheric carbon dioxide. All purified water will readily pick up this gas at an air interface, especially if it is sprayed. This produces carbonic acid, which will react with metallic oxides to produce insoluble or near-insoluble basic carbonates [40]:

$$Co_2 + H_2O = H_2CO_3$$

$$2MeO + H_2CO_3 = MeCO_3Me(OH)_2$$

These basic carbonates, present on metallic surfaces, are ionogens and may decompose readily into ionic contaminants in the presence of most acids and alkalis [40]. For example:

$$MeCO_3Me(OH)_2 + 4HCl = 2MeCl_2 + CO_2 + 3H_2O$$

Since this reaction series is self-sustaining, it can be dangerous. It is most easily avoided by initially having a minimum quantity of metallic oxide on the parts.

A more detailed diagram of contaminants from cleaning processes is shown in Figure 9.6. For classification of the contaminants, please refer back to Table 1 (pg. x) for each by-product.

9.4.6 Conformal Coating

Process residues encapsulated in the conformal coating can deteriorate the reliability of circuits because they increase the possibility of corrosion, vesication, blistering, electromigration, and other stresses affecting long-term coating adhesion (this virtually eliminates no clean/LSF from use on circuits that require coatings). Boards must be completely clean to prevent process residues from being trapped under the coating and from causing reliability problems in the future (for more information, see the chapter on conformal coatings).

9.4.7 Testing

Dry testing of high voltage circuits may be performed with the use of sulfur hexafluoride and its mixtures with nitrogen. Both of these gases, under normal conditions, are nearly chemically inert and without contamination. However, at certain dielectric stress levels, such as when the applied voltage stress exceeds the dielectric strength of the gas, sulfur hexafluoride decomposes [44]. Its decomposition products are the lower sulfur fluorides (tetra fluoride or difluoride) and fluorine gas:

$$SF_6 = SF_4 + F_2 \qquad \text{or} \qquad SF_6 = SF_2 + 2F_2$$

In the presence of moisture, hydrolysis can cause the formation of corrosive compounds such as hydrogen fluoride and oxyfluorides:

$$SF_4 + F_2 + 2H_2O = 4HF + SO_2F_2$$

As a result, corrosion products can form on a variety of metals due to the interaction of the corrosive decomposition products with the metals [44].

Nitrogen, under arcing conditions and in the presence of oxygen, can form oxides that function as anhydrides of strong acids, such as N_2O, NO, NO_2, N_2O_3, or N_2O_5. In the presence of moisture, the hydrides will form acids, and the effects are similar to those described for fluorine compounds. Appropriate contamination control prevents oxygen and moisture from being present in the gas or gas mixture and involves frequent changes of the test medium [44]. Liquid dielectrics that may be used are fluorinated ethers, chlorofluoro hydrocarbons, mineral oil, and polybutene. The use of silicone fluids is unwise if the assembly or module tested has to be encapsulated; quantitative removal of silicone fluids is nearly impossible [44].

Contamination from ovens and environmental chambers can occur. Sources encountered are typically hot air, dust and dirt, carbonic acid, and contaminated water from humidity chambers. Carbon dioxide is used for cooling environmental chambers, since the cooling cost is considerably lower for CO_2 than for N_2 [44]. Carbon dioxide is the anhydride of carbonic acid, readily combining with water in the equation:

$$CO_2 + H_2O = H_2CO_3$$

Carbonic acid is a weak and instable acid, but it is ionically very active, providing a good electrolyte for the initiation of corrosion and dendritic growth. Unprotected metals and mismatched metal pairs are affected when contacted or bridged by water films containing dissolved CO_2. Preventive measures include minimizing moisture in the chambers and protecting the metals [44].

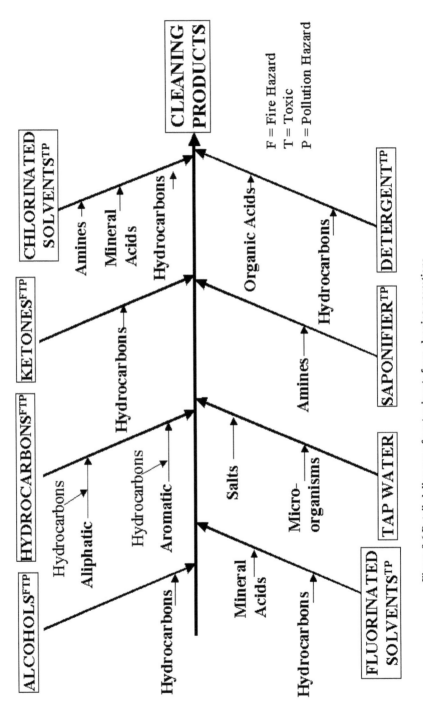

Figure 9.6 Detailed diagram of contaminants from cleaning operations

9.4.8 Repair and Rework

After cleaning, assemblies are usually inspected, tested, and retouched. In addition to contamination from handling, contamination can be introduced due to desoldering and hand soldering.

Two techniques are generally used for desoldering [46]. Aspiration of the molten solder with a special soldering iron generally involves noncontaminating solder oxidation and is not problematic. Aspiration by a separate spring-loaded pump is also fairly clean, provided the teflon tip is replaced as soon as it shows the slightest discoloration (thermal decomposition of teflon will occur at soldering iron temperatures, liberating small quantities of hydrofluoric acid).

Solder removal by desoldering tresses is far from harmless [46]. Tresses are woven from fine copper wire, sometimes tinned, and then cleaned and fluxed. The fluxes used are sometimes relatively mild, but can be very strongly activated. Any board that has had a tress used on it should be carefully cleaned. The fluxes impregnating these tresses are in contact with the air and deterioration is fairly rapid. An old tress can contain an oxidized flux, which may be difficult to eliminate, even though it may still wick the solder satisfactorily. This method of solder removal, while valid, may introduce fluxes whose type and composition may be unknown.

Although hand-soldering techniques are similar to machine soldering using resin fluxes, there nevertheless remain several differences involving temperature and the flux composition. Mass soldering techniques use temperatures from 230℃ to 260℃, usually in the middle of this range. Hand soldering may involve temperatures from 250℃ up to 450℃, but most frequently within the range of 330℃ to 390℃ [47]. The effect of these higher temperatures is to modify the nature of the residues. Carbonization (charring) is much more pronounced, rosin is chemically modified, organic acid activators are more fully polymerized, and halide activators are more fully decomposed. The flux itself will probably contain more plasticizers to prevent it from breaking up when the solder wire is flexed.

The result is that the chemistry of flux removal will differ from that for machine soldering flux. Optimum removal may seldom be obtained for both types of residue with one machine containing one solvent. In most cases, two separate installations with different processes are necessary.

Users of water-soluble fluxes often want to obtain a solder wire with a similar flux incorporated for retouching, so they can use the same equipment for cleaning after rework. Some water-soluble flux wires are unsuitable for fine electronics soldering, either because their residues are hard to eliminate or because they contain inorganic fluxes based on ammonium chloride [47]. The easy way around this problem is to use a solid solder wire, dipping the end from time to time in an "ink pot" containing the same flux as used for soldering.

To keep the residues soluble when using soldering irons with a water soluble flux, use the lowest possible temperature, preferably less than 300℃ [47]. This is feasible since the high flux activity ensures a faster solder flow than do other types.

When using relatively low iron temperatures, it is best to use the largest possible bit size compatible with the dimensions of the solder joint to minimize the drop in temperature. Using a low temperature will also reduce the bit wear.

Another problem with repairing and reworking a component is that it can introduce excessive heating of the circuits. This can lead to localized decomposition of polyurethane coatings, which break down into carbon dioxide and water when heated. Thermoplastic cases can crack and melt due to thermal stresses. Circuits near the source of heat can break and change the dielectric. Finally, thermoplastic acrylics can decompose and leave residues on the circuitry [86]. Components that are reworked are often not cleaned using the same methods as in assembly and are thus more likely to be contaminated [51]. As a result, users must be careful not to heat the components more than necessary during rework. They must also be sure to use clean equipment and handle the components correctly.

9.5 Use Conditions

During its service life, an electronic product is continuously exposed to conditions that can cause contamination. The potential for failure due to the contaminants encountered depend on the sensitivity of the circuitry, the materials and components used, the applied safeguards, and the duration and severity of exposure.

The field environment is a source of a great variety of contaminates that adversely affect assemblies cleaned of process contaminates but inadequately protected. Airborne contaminants pose a serious threat to electronic systems because of the complex and complicated synergistic chemistry involved, as they generate numerous ionic and hygroscopic microscopic compounds. These pollutants exist in three chemical stages. The first is the generation of very reactive gaseous agents (NO_x, SO_x, O_3, H_2O_2, NH_3, H_2S, plus VOC's (Volatile Organic Compounds)). The next stage is suspended inorganic and organic acids (H_2SO_4, HNO_3, HCl, $HCOOH$). The final stage is suspended submicron hygroscopic ionic particles (NH_4HSO_4, NH_4NO_3, $[NH_4]_2SO_4$) plus numerous others. Additionally, there are many other inorganic compounds and metals, such as sodium chloride, iron, aluminum, magnesium, calcium and potassium, that are suspended in the air.

Unfortunately, all three stages can be present at any given time. Which stage is predominant depends on the time of year, temperature, humidity, local weather conditions, effectiveness of environmental control methods, and the length of time the circuitry has been in use. All three stages require the presence of moisture that will cause the chemical reactions necessary to degrade electronic reliability.

The following table shows the amounts of particles in a cubic foot of air, the mass of these particles per cubic meter, and the rate of deposition of these particles per square foot. It should be noted that the use of internal forced air cooling with an outside air source can increase the rates of particle deposition by up to 100 times. For more information, see section 9.1.4 Airborne Particles.

Additionally, salts, dust, and fungus can also result in the degradation of electronics. Their occurrence and effects are enhanced by temperature and humidity, such as encountered in tropical climates. Fungi can damage or completely destroy many organic materials and some minerals by excreting enzymes that exert a digestive action on the substrate. Fungicidal agents, followed by aqueous rinsing and drying, can remove most fungal contaminates. Contamination control requires not only an adequate design for the anticipated environment, but also regular servicing.

During its operating life, an electronic product is also subject to contamination as the components degrade as a result of inadequate design, thermal runaway, or electrolyte leakage. Inadequate material design can include the use of substrates that may produce hot spots in excess of the rated maximum temperature tolerance of the laminate [13]. The result will be a pyrolytic degradation of the epoxy resin contained in the laminate. The polymer chain of the epoxy will break and reduce the electrical insulating characteristics. The insulator is converted into a semiconductor and eventually a conductor and has thus become a contaminant. This can happen when the materials have a different thermal output than that of the components or PWA. When a component overheats because of electrical overstress, components malfunction and overheat, allowing shorts and current leakage in the capacitors or diodes. Eventually, solder masks can vaporize, if the temperature gets above 125°C, creating worse problems [86].

Another potential contamination source is electrolyte leakage from some capacitors. Internal component malfunction, the shape of the applied electrical signal, or incorrect component installation can cause overheating. Poor seals on capacitors can lead to leakage causing corrosion. The buildup of vapor pressure from the electrolyte inside the sealed component may break the seal; the component can literally explode, causing corrosive electrolyte spillage over the assembly.

TABLE 9.9 Particle concentrations and rates of deposition

Adolescent Particles (0.01- 0.1∝m)			
Concentration Particles/ft^3	Outdoors 10^8-10^{10}	Office Building 10^4-10^7**	Cl-10 clean room 10-500
Rate of Particle Deposition ft^2/sec (Average)	Outdoors 2×10^6 - 2×10^9 /ft^2/sec	Indoors 3 - 2×10^5 /ft^2/sec	
Mature Particles (0.1-1.0∝m)			
Concentration Particles/ft^3	Outdoors 10^5 - 10^8	Office building 10^6 - 10^7	Cl-10 clean room 1 - 10
Rate of Particle Deposition ft^2/sec (Average)	Outdoors 2×10^3 - 2×10^6 /ft^2/sec	Indoors 200 – 2000 /ft^2/sec	
Minerals Biologicals (1-15∝m)			
Concentration Particles/ft^3	Outdoors 10^3 - 10^5	Office building 10 - 10^3	Cl-10 clean room 0.5 - 1
Rate of Particle Deposition ft^2/sec (Average)	Outdoors 20 – 2000 /ft^2/sec	Indoors 0.2 – 20 /ft^2/sec	

CHAPTER 10 MEASURING CONTAMINANTS IN PRINTED WIRING ASSEMBLIES

T.F. Egan at the Bell Laboratory started cleanliness testing in the early 1970s. Egan employed a conductivity meter and made aqueous extracts to investigate the effect of plating salt residues on specific conductance. Today there are various industry-accepted techniques for determining cleanliness levels after the PWA has been defluxed.

10.1 Testing for Ionics

After cleaning, the effectiveness of cleaning materials and residue removal processes should be evaluated. As we discussed, flux residues may contain complex mixtures of ionic and nonionic materials. The ionic portion of these residues is usually easier to detect because its effects on the electrical conductivity of the solution indicate ions in water or water/alcohol solutions. Because both ionic and nonionic materials are deposited on the board surfaces together, measuring the ionic portion can serve as a convenient tracer for residual flux residues.

In order to measure ionic residues, they must be extracted into an aqueous solution. The presence of ions in water can then be measured quantitatively by measuring its conductivity. Because flux residues often contain components that are not readily water soluble, mixtures of water and an alcohol, such as isopropanol, are usually preferred as the extracting solvent. This mixture dissolves the water-soluble ionic constituents, as well as the alcohol-soluble nonpolar constituents, allowing the measurement of any ionic materials that may have been embedded in the nonpolar residue. The industry standard for the amount of ionic contamination at the substrate and assembly levels is 1.56 μg/cm^2. Testing for ionic contaminants is of greater concern than testing for nonionic contaminants, since ionics tend to be more detrimental.

10.1.1 Ionic Contamination Test

The ionic contamination test uses a simple conductivity cell to measure the specific conductance of a test extract. R.J. DeNoon and W.T. Hobson at the Navy Avionics Facility determined that a test extract employing 75 vol.% isopropanol (IPA) and 25 vol.% water (H_2O) was the most suitable for dissolving rosin-based flux residues. This test, using a conductivity meter, has remained standard up to the present day [13].

This cleanliness test became incorporated in U.S. MIL-P-28809, the specification on the acceptability of U.S. military PWB assemblies. Soon after, several companies brought out an automated version of this test, since the manual method is tedious and requires extreme care. Today, these automated machines have programs stored so that the operator has only to enter the dimensions of the assembly and adjust the volume of the test solution [13].

Although the solvent extract resistivity test (also called the ionic contamination test) proved relatively easy to implement, it suffers from several drawbacks. It does not distinguish among different ionic species, and does not detect the presence of nonionic and nonionizable (but polar) species, even if they are soluble in the test medium [13].

Some work has been performed using direct surface analysis techniques, but these methods are generally too costly, both in equipment and personnel, to be used routinely. Thus, extractive techniques are the method of choice for cleanliness testing. A great deal of progress has been made in employing extractive techniques for cleanliness testing, but what is required is an industry consensus that one or several of these techniques is beneficial and its subsequent incorporation into an acceptable industry standard [13].

Several non-extractive techniques are available. These are based on the application of a bias voltage across a special trace pattern while the PWA is subjected to either moisture or moisture/temperature cycling. These tests do not quantify the amount of contamination that is present, but rather monitor the resistance of the test pattern. Sharp drops in resistance indicate an event is occurring. Some events, such as dendritic growth, are contamination related. Surface insulation resistance (SIR) testing and electromigration (EM) testing fall under this category.

10.1.2 Ion Chromatography

Although the ionic contamination test fulfills a definite industry need, it does not and cannot distinguish among different ionic species. The capability to distinguish among different ionic species and to quantify the amount of each provides much more effective process control. Different ionic species come from different parts of the fabrication and assembly operations, and the ability to distinguish different species would enhance the ability to control the entire operation more effectively.

Ion chromatography can detect and quantify anions, such as chloride, bromide, and sulfate, and cations, such as sodium, potassium, and calcium [4].

To detect and quantify the amount of each ionic species, ion chromatography can be used. As in all chromatographic methods, columns are used to separate a given mixture so that the constituents exit at different times (the elution time); a detector is used to identify the species. The detector signal is converted into an analog signal, which appears as a curve on graph paper. The area under this curve corresponds to the area of a standard, allowing the amount of the species to be quantified. Each species has its own characteristic elution time. Today, many chromatographs also have direct digital readouts [4].

10.1.3 Omega Meter and Ionograph

The Omega Meter uses static-volume extraction method similar to the ionic contamination test, in which a sample is immersed in a fixed volume of an alcohol/water mixture. After the extraction, the resistivity of the solution is read. This is expressed as an equivalent amount of an ionic salt, such as sodium chloride, per unit area, representing the total amount of ionic material removed in the extraction [55].

The ionograph uses a dynamic system of ionic extraction similar to IPC-TM-650 test method 2.3.26. In this method, the extracting water/alcohol solution is pumped continuously in a closed loop from a sample tank through a conductivity cell and an ion exchange column, and then back to the sample tank. When the sample is immersed in the tank, the conductivity rises to a peak, and then gradually falls back to the pre-immersion level [55]. Measurement and integration of the conductivity over the time of extraction gives values that are linear functions of the amount of ions taken in solution.

10.2 Testing for Nonionics

Tests for nonionic contamination are performed with certain types of machinery; some identify the contaminants, others do not. The methods are discussed below.

10.2.1 High-performance Liquid Chromatography

A method to detect nonionic species on PWAs is called high-performance liquid chromatography (HPLC). This method will detect and quantify the residue. Moreover, if peak identification can be made, HPLC is species-specific; for example, it will detect and quantify the amount of a single species, such as abietic acid [13].

Like any chromatography system, it involves the use of an extractive medium (generally acetonitrile), an injection system, columns on which different materials are selectively absorbed so that they elute out of the column at different times, and some sort of detection device. For nonionic residues derived from rosin fluxes and pastes, a UV/VIS (ultraviolet/visible) detector senses any species absorbing light in the UV or visible range of the spectrum (180 nm to about 700 nm). The HPLC technique is nondestructive, but the equipment needed is relatively expensive [13].

HPLC is very effective in analyzing white residue (WR) resulting from thermal degradation flux residues. Several investigations using HPLC strongly implicate metal abietates of tin and lead as the chief promoters of WR. An abietate is the salt formed between a metal and any rosin acid. Tin/lead abietates form during flux/soldering. When the activator acts on the tin/lead oxides, tin/lead salts of both rosin and chloride ion–termed monoabietate metal salts–are probably formed [13]. In the presence of excessive heat and moisture, these salts dimerize (the molecule expands to approximately twice its size) to form diabietate and didehydroabietate metal salts, both of which are much less soluble than monoabietate salts. Diabietate and didehydroabietate metal salts are formed during the soldering process; however, the white residue normally appears after defluxing, since many solvents do not have enough solvating power to remove these tenacious species [13].

10.2.2 Ultraviolet/Visible Spectrophotometry

Ultraviolet/visible spectrophotometry is another useful technique for detecting and quantifying contaminant species. Extractions from PWAs are made using isopropanol (reagent grade or better). Extraction tanks made from stainless steel are the most suitable for this purpose, but plastic bags have been used. It is important to calibrate the instrument using a flux to produce the calibration curve, preferably the actual flux or paste whose residues are being determined. Rosin acid residues typically absorb between 200 and 250 nm [52]; values of rosin residual typically run between 200 and 600 $\mu g/in^2$. The test is nondestructive.

10.2.3 Coulometry

Coulometry is a rapid, simple, reproducible, and quantitative method that has proven successful in detecting organic residues. Coulometry involves the combustion of sample carbon and its conversion into carbon dioxide (CO_2) inside a tube furnace held at 400 to 500°C. Once the carbon is converted and dried, the carbon dioxide is reacted with ethanolamine to form carbamic acid [$HOCH_2CH_2NHCOOH$], which is titrated coulometrically. Since this method involves the pyrolysis of organic material and its conversion into carbon dioxide, it should not be used with any materials containing organic matter, such as laminates and components [55].

This method is applicable with ceramic carriers; however, the method quantifies only the amount of carbonaceous residue remaining on the ceramic component, and therefore is not species-specific. Since ceramic components must be removed to perform the test, the test is destructive. It is very effective, however, in quantifying the amount of organic residue and can be used as a process control and parameter aid. Results are typically expressed in micrograms of carbonaceous material per square inch (μg C/in^2) [55].

10.2.4 Turbidimetry

Turbidimetry can be used to detect and quantify nonionic (principally rosin) residues on PWAs [55]. It is an extractive technique using isopropanol (reagent grade or better). The turbidimeter must be calibrated using standard solutions. Once extracts are made, hydrochloride acid (0.5N) is added to turn it cloudy. The cloudier the extract, the more rosin residue it contains. The results are given in micrograms of rosin per square inch (μg rosin/in^2).

10.3 Surface Insulation Resistance Test

Among the few nonextractive test methods for ascertaining cleanliness of PWAs, surface insulation resistance (SIR) testing is probably the most common. SIR determines the resistance between parallel traces of a Y pattern or a comb pattern on bare boards. Such a resistance will normally be very high ($\sim 10^{11}$-10^{12} ohms) unless something on the surface causes it to drop. Thus, the test can function as a method for checking surface contamination [46].

The SIR test is typically conducted at elevated temperature and humidity over an extended time period. The test patterns are normally exposed to a bias voltage (excitation voltage), followed by a measurement voltage. The excitation voltage and measurement voltage may or may not be the same polarity. Certain types of contamination can, with elevated temperature, humidity, and excitation voltage, lead to leakage currents across the conductor traces of the test pattern and a lowering of the SIR [5].

There is no one industry-accepted test pattern for SIR testing. Examples of test patterns are found in the IPC-B-25 test board and the Bellcore test pattern [145]. To relate different test patterns to each other, all resistance readings must be expressed in ohms per square. The number of squares is found by dividing the length of the line in inches by the spacing width in inches and then multiplying by the number of spaces. Table 10.1 shows a comparison of various SIR patterns. Nevertheless, even if SIR data are reported in ohms per square, a standard test pattern and comparable excitation and measurement voltages should be used [46].

Table 10.1 Comparison of various SIR test patterns

Pattern Type	# of Spaces	Line Length (in.)	Space Width (in.)	# of Squares	Minimum Resistance Value (x 10^9)
One Square	1	1.0	1.0	1.0	337.0
Bellcore	5	1.125	0.050	112.5	3.0
IPC 100043, J1	17	1.0	0.020	850.0	0.4
IPC 100043, J2	17	1.0	0.010	1,700.0	0.02
IPC B-25, A	23	0.625	0.00625	2,300.0	0.015
IPC B-25, B	11	0.625	0.0125	550.0	0.61
IPC B-25, C	5	0.625	0.025	125.0	2.7

The lack of a standard pattern and voltage is one drawback of SIR. Another is the length of time it takes to perform the test. A seven- to ten-day test is typical, and some SIR tests extend for six weeks. Finally, although SIR testing will reveal the presence of contamination that causes leakage currents, it is not species specific. It cannot, by itself, identify whether the contamination is ionic or nonionic.

10.4 Electromigration Test

The Electromigration (EM) test involves placing a bias voltage across two conductor traces and determining the time to failure. The time to failure is generally found by picking a threshold value for the current (usually 500 μA) and measuring the time it takes to reach that leakage current [50]. Some type of moisture exposure is generally required to conduct the test. Under the proper bias voltage, dendritic growth takes place between the traces, causing a rise in current and a breakdown in the resistance between the traces. Typically, a bias voltage less than 10 volts is best for producing dendritic growth. Sometimes the test is run in a temperature/humidity chamber with the same pattern used for SIR testing; a drop of deionized water may be used under ambient conditions to produce the phenomenon [50].

The EM test does indicate the presence of contamination on the surface that can lead to leakage currents and dendritic growth. Also, if the proper test pattern and bias voltage are used, the test can be performed quickly. The chief drawback of the test is lack of standardization. At present, there is no standard test pattern, voltage, or methodology for performing the test. Also, the method is not species-specific [55].

CHAPTER 11   QUALITY AND RELIABILITY OF PRINTED WIRING ASSEMBLIES

Defects can be generated during assembly. In addition, as with laminate and substrate manufacturing, some defects discussed previously can also be precipitated during assembly operation. Sometimes these defects are not detected at the earlier stages of assembly, but mature, becoming observable through the many steps required to assemble a board. These defects can impact overall reliability, because they can randomly occur and grow, negatively affecting the quality of the assembly, and reducing reliability.

11.1 Flux Residues

A visible residue from cleanable fluxes or activated flux residues on electrical contact surfaces (Figure 11.1) are considered possible nonconforming defects. Flux residues are directly correlated with the type of flux and type of cleaning used. If residues are not adequately removed, they can become entrapped under components, where they could expand and add stress to the solder joint, initiating a crack. If residues are entrapped under a conformal coating, they will reduce the adhesion of the coating to the board.

11.2 Particulate Matter

Particulate matter can be detrimental on an assembly because it can absorb moisture and provide a conductive path between conductors, causing an electrical short (Figure 11.2). Particulate matter includes dust, dirt, lint, dross, dander, and hair. Assemblies must be free from these particles especially if a conformal coating is to be applied.

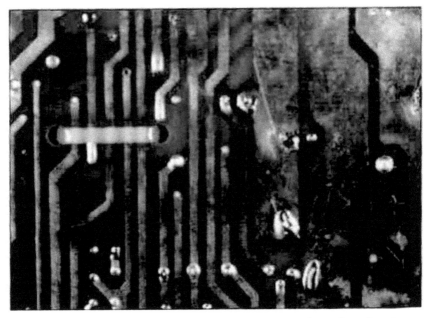

Figure 11.1 Photograph of flux residue on the surface of a printed
wiring board [73]. With permission from Institute for Interconnecting and
Packaging Electronic Circuits.

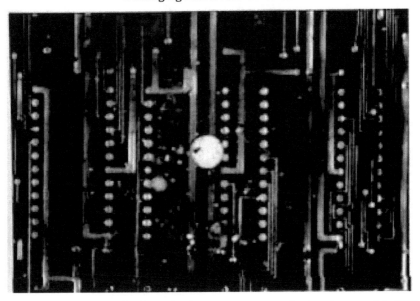

Figure 11.2 Photograph of particulate matter on an electronic assembly [73].
With permission from Institute for Interconnecting and Packaging Electronic
Circuits.

11.3 White Residues

White or gray areas of residue on a PWB assembly surface (Figure 11.3), which occur after soldering and cleaning, can affect many production operations. White residue can be of several types, but other phenomena can give the illusion of white residue when in fact there is no residue at all. For example, an incompletely cured solder mask, or one that is chemically incompatible with the cleaning agent used on the assembly, may leave a whitish residue.

The most common cause of white residue is oxidation of rosin. Rosin flux constituents oxidize due to excessive solder temperature, flux aging, or moisture when the rosin-fluxed assemblies are cleaned with organic solvents. Polymerization of small amounts of rosin from heat exposure may also play a role in the formation of white residue.

Lead chloride is another common cause of white residue. Lead chloride, generated by the use of hot chlorinated water or of 1,1,1 trichloroethane containing hydrochloric acid, will be most pronounced on and adjacent to solder-coated conductor areas, as opposed to the entire board surface. Lead chloride growth, once initiated, continues with action from moisture and carbon dioxide. The resulting corrosion produces lead chloride and lead carbonate on the affected conductor. Lead chloride has ionic activity and is capable of producing electrical effects. Overcoating the lead chloride with a conformal coating will produce vesication when the assembly is exposed to humidity.

Reservoirs containing cleaning media become cumulatively contaminated with flux products, assembly construction materials, and particulate debris from the assembly. When a rinsing procedure does not remove these contaminants, they are deposited when the cleaning liquid evaporates. A surface discoloration caused by these deposits often resembles white residue. The effects of these residues are unpredictable, due to the variety of materials that may be encountered.

Table 11.1 Commonly Formed Metal Salts After Soldering. With permission from CALCE Electronic Products and Systems Center.

Compound	Water Solubility (g/100cc)	Color
$CuCl_2$	70.6	Green
$CuCl$	.006	Green
$CuBr_2$	Very soluble	Black
$CuBr$	Slightly soluble	White
$SnCl_2$	83.9	White
$SnBr_2$	85.2	Pale yellow
$PbCl_2$	1.0	White
$PbBr_2$	0.8	White
$PbCO_3$	Insoluble	White
$CuCO_3*Cu(OH)_2$	Insoluble	Green
Copper resinate	Insoluble	Green
Tin resinate	Insoluble	Tan

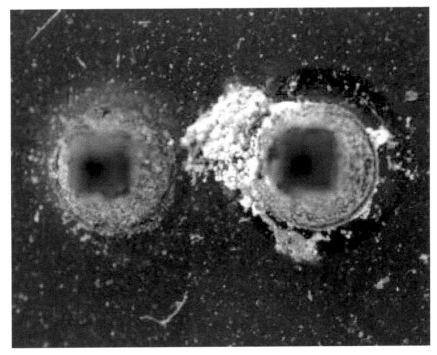

Figure 11.3 Close-up of an area of white residue potentially connecting two conductors. With permission from CALCE Electronic Products and Systems Center.

11.4 Solder Voids

Solder voids occur due to contamination of the solder or the bonding surface. On a typical plated through-hole, voids can be located in the solder itself or at the solder/copper interface (Figure 11.4). If the voids are large enough, they look like poor wetting at the interface.

11.5 Corrosion

Even though corrosion is usually viewed as a reliability problem because it affects the performance of the assembly over time, it can also be viewed as a quality problem. Corrosive areas can be observed during or very shortly after the assembly process due to the environment and cleaning chemicals. However, there is a chance that existing corrosion may not affect the reliability of the product, if the corrosion forms thin insulating layers on the contact interfaces. Corrosion can arise even on noble plated metals if there are defects. For example, corrosion can occur in the pores of a gold layer over nickel and copper. Figure 11.5 depicts various corrosion examples.

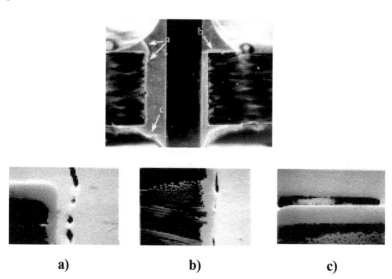

a)   b)   c)

Figure 11.4 The top photograph shows three locations of solder voids; photos a, b, and c are close-ups of the areas. Photo a) shows voids in the solder, photo b) shows voids at the solder copper interface, and photo c) shows larger voids, which is interpreted as poor wetting at the interface. With permission from CALCE Electronic Products and Systems Center.

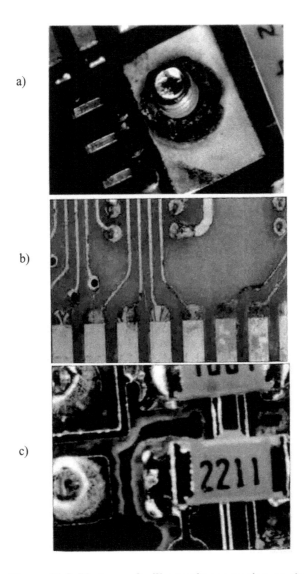

Figure 11.5 Photographs illustrating corrosion on a) a bolt connector, b) traces and pads, and c) around bond pads [73]. With permission from the Institute for Interconnecting and Packaging Electronic Circuits.

11.6 Coating Deadhesion

Deadhesion of conformal coatings or permanent solder mask is a defect. Like the solder resist defects discussed earlier, coating deadhesion can lead to reliability problems due to moisture impregnation generating corrosion, electromigration, leakage currents, and so on. Defects include any voids, bubbles, dewetting, ripples, fisheyes, or foreign material that expose or bridge circuitry, lands, or adjacent conductive surfaces (Figure 11.6).

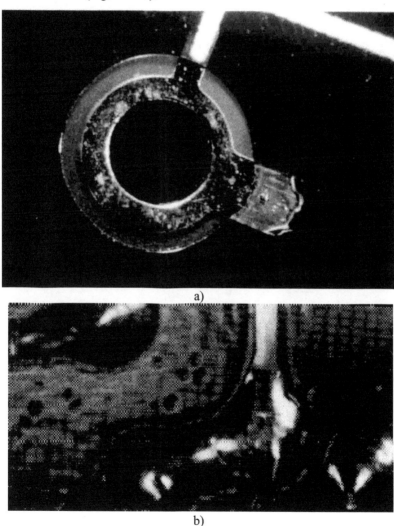

a)

b)

Figure 11.6 Coating defects including a) voids and blisters and b) fisheyes [73]. **With permission from the Institute for Interconnecting and Packaging Electronic Circuits.**

Figure 11.7 Coating defects including a) cracking b) wrinkling [73]. With permission from Ingemar Hernefjord from Ericsson MicroWave.

11.7 Black Pad

An intermetallic layer which forms during soldering will result in an embrittled solder joint, leaving a smooth black or dark surface, see Figure 11.8. This is sometimes referred to as black pad. This is caused by contamination of the electroless nickel plating baths. The chemical bath for the electroless plating of nickel requires constant monitoring to prevent build-ups of contaminants and to insure the proper chemical balance [146].

Copper substrates prepared by pumice grinding that are insufficiently cleaned may cause the nickel plating to be porous. This may be magnified if the nickel plating bath is contaminated. Porous plating provides an avenue that the copper can diffuse through, which may embrittle the solder joint as well [146].

Figure 11.8 Failed BGA solder joint due to "Black Pad." With permission from Ingemar Hernefjord, Ericsson MicroWave.

CHAPTER 12   CONTAMINATION REMOVAL

The terms *contamination removal* and *decontamination* describe the removal of any matter that may be considered a contaminant. Decontamination can be separated into the broad categories of cleaning (with solid, liquid, or gaseous media), and drying (through volatilization). The objectives of contamination removal are to remove contaminants cost-effectively and efficiently satisfying the applicable quality standards and requirements for processing a specific product volume over a given time.

The expense of contamination removal is justified when the removal prevents detrimental effects in production or use. Contamination removal must be timely. It must occur as soon as possible after contamination introduction because aging may change the removal characteristics of the contaminants.

Abrasive blasting with small particles is an effective way to remove surface contaminants without regard to their solubility. This technique removes not only insoluble contaminants, but also contaminants included in the surface being cleaned. Since abrasive blasting enlarges the surface morphology by imparting roughness, polymeric materials bond better with the treated surface, yielding better environmental protection. On the other hand, if a coating does not protect the roughened surface, the roughness promotes the formation of water films and, in the case of metallic surfaces, an increased tendency to corrosion.

Fluids remove surface contaminants by dissolving the matter, float-off effects, and/or cavitation effects. The rate and efficiency of contamination removal depends on the solubility of the substance removed and the dissolving capability of the liquid. The amount of time the solution is in contact with the matter, the amount of agitation during the process, and the temperature of the solution affect the efficiency of the removal. Usually, the solubility increases as the temperature increases due to decreased surface tension, but there are exceptions, such as amino acids contained in fingerprint deposits.

Surface tension is an important property of a liquid cleaning medium, especially when high-pressure spray cannot reach shielded areas. Low-surface-tension fluids are preferred for cleaning; surface tension is lowered by heating the liquid or by adding surfactants to the wash solutions. Surfactants or wetting agents lower the surface tension of the fluid when added in relatively small amounts. Surface tension values of a variety of fluids used for electronic production cleaning are shown for comparison in Table 12.1.

Table 12.1 Surface tension of common cleaning liquids [4]. With permission from C. J. Tautscher.

Liquid	Temperature (°C)	Surface Tension (dyn-cm)
Water, pure	5	74.90
	20	72.75
	40	69.56
	80	62.60
6% saponifier, aqueous medium	20	30-40
Water, pure, surface tension attainable with fluorocarbon surfactants	20	17-35
Water, pure, with 0.005% nonionic surfactant, Rohm & Haas DF-18	20	34.00
Toluene	20	28.50
Methylene chloride	20	24.60
Methyl ethyl ketone	20	24.60
Ethyl alcohol	20	22.75
	30	21.89
Methyl alcohol	20	22.61
Isopropyl alcohol	20	21.70
CFC-1 azeotrope with methylene chloride	20	21.40
n-Hexane	20	18.43
Trichloro trifluoro ethane (CFC-1)	20	17.30
CFC-1/methyl alcohol blend	20	17.40

Removal by float-off, a side effect of removing contamination by dissolution, occurs when a contaminating deposit consists of two or more contaminants with different solubility characteristics. A good example is fully activated rosin flux, which consists of water-insoluble rosin and a water-soluble activator. The activator is typically dispersed throughout the rosin deposit after completion of soldering. Rosin removal with 1,1,1 trichloroethane will dissolve the rosin and free the amine activator; the water-soluble amine will float off as the surrounding rosin is being removed. Hence, the insoluble rosin removes the bulk of the activator.

The scrubbing action of ultrasonic cavitation produces contamination removal that is similar to float-off. The formation of numerous bubbles and their implosions from ultrasonic frequency vibrations causes the dislocation of contamination deposits. When the deposits are soluble in the liquid medium used, their solubility is enhanced. When the deposits are insoluble, the particles will float off in the liquid medium.

Various gases are used for contamination removal. Two major applications are gross drying and plasma cleaning. In the former, compressed air is commonly used to blow off surface water after aqueous cleaning or between aqueous process steps. The time required for "fine drying" becomes significantly reduced, allowing adequate drying for many commercial products with a brief in-line heat drying. This is typically accomplished with air-knives, whereby the compressed air is released on assembly surfaces at the optimal angle to force the jet stream under the components, blowing the water film off. Compressors are used to pressurize the air, which can raise a concern about oil contamination if subsequent coating is to be performed without additional cleaning. Dry and pure compressed nitrogen is used for critical applications [4].

In plasma cleaning, gases such as oxygen, argon, and tetrafluoromethane (CF_4) are employed by exposing their atoms to radio frequency radiation. The resulting atomic excitement causes high-velocity atoms from the gas to bombard surfaces in the plasma chamber. Organic molecules impacted by the gas atoms are chemically changed. Polymer molecules break up into shorter chains, and when oxygen is used, chemical oxidation assists the break-up action. This, combined with the resulting temperature increase from the radiation, rapidly vaporizes small amounts of organic compounds. Volatilization products are vented through a vacuum system [49].

12.1 Abrasive Cleaning

Solid abrasive materials are generally applied with air pressure and require an oil-free air supply. Pressure accelerates the abrasive action; experimenting first with expendable parts can help determine the optimum pressure. Masking is required to shield areas such as markings or other surfaces that need to be maintained in their original state. Abrasive blasting may also cause buildup of static charges, so precautions must be taken to protect static-sensitive devices by effective grounding [18].

Some common media for abrasive blasting are listed in increasing order of abrasiveness [18]:

Ground walnut shells are used on softer materials. Since walnut shells contain oils, either "deoiled" shells must be used or the abrasively cleaned item must be degreased after the treatment to avoid an oily film on the cleaned surfaces. Walnut shells are typically used dry. The usual particle size is 250 microns.

Sodium bicarbonate can be used dry or in the form of a wet slurry, on which the saturated bicarbonate solution contains an excess of this salt. It has been used for selective removal of conformal coating films, but embedded particles of the strong ionic bicarbonate must be avoided. Mild acid rinses, a hot water soak, and a thorough rinse with deionized water are recommended post treatments. Precautions must be taken against corrosion, especially when unprotected galvanic couples are part of the cleaned item. A typical particle size is 100 microns.

Glass beads, used for harder surfaces, are not recommended for removal of conformal coatings, solder finishes, and similar soft surfaces. A typical particle size is 50 to 100 microns.

Silicon carbide is commonly used for hard surfaces, such as steel. Typical particle size is also 50 to 100 microns.

Since abrasive cleaning is a messy operation, precautions must be taken to prevent dust and other debris from entering other process areas. Frequent vacuuming of the abrasive blasting area is recommended.

The abrasive wheel brush uses a nonfibrous nylon material containing silicon carbide. It has several purposes in the preparatory stage of the PWB fabrication process: (1) deburring action on the drilled holes prior to plate-through; (2) decontamination of copper cladding surfaces by removing surface oxidation, organic soils, and fingerprint deposits [18]; (3) provision of a contamination-free surface for metallization and resist applications.

In abrasive wheel cleaning, a solid agent removes the surface contamination via mechanical scrubbing and abrasion, a liquid removes loose particles, and compressed gas applies forced drying, blowing off surface water. The abrasive wheel brush treatment is superior to pumice brush cleaning because it does not leave abrasive particles embedded in the surfaces.

The nonabrasive bristle brush treatment is confined to cleaning bare circuit boards, usually before solder mask application or as a final step after board manufacture, to complement the abrasive wheel [18]. Increased demand for a higher level of board cleanliness makes the use of these bristle brushes very desirable. An aqueous detergent solution is applied with a gentle brushing action from soft bristles and followed by a hot purified water rinse provides effective contamination removal.

12.2 Solvent Cleaning

Solvent cleaning is on the decline because of concerns about the solvent's impact on the environment. New solvents have been developed but their use may also be curtailed because of the same concerns. Until recently, solvent cleaning has been the method of choice to remove the rosin-based fluxes common in solder paste.

Solvent cleaning features very short drying times while producing no effluent to be drained. The basic fundamentals of solvent cleaning are the same whether a batch or in-line system is used [74]. When a board assembly enters the system, hot vapor condenses on it, beginning the cleaning process. The assembly is then sprayed with solvent or completely immersed in solvent.

Solvent cleaning was the most popular cleaning process before the Montreal Protocol agreement. The subsequent phase-out of chlorinated fluorocarbons (CFCs) and other ozone-depleting substances have curtailed the use of the process, but there are other solvent-based cleaning options. Isopropyl alcohol (IPA), hydrofluorocarbon (HFC), hydrochlorofluorocarbon (HCFC), perfluorocarbon (PFC), and hydrofluoroether (HFE) seem to be the most widely used solvents for electronics. IPA is highly flammable but special equipment handles it [74]. PFCs, HFCs, and HFEs are inefficient cleaners by themselves, but can be effective when combined with another solvent in a co-solvent process. In co-solvent cleaning, two immiscible agents remove board residues. One acts as the cleaner, and the other displaces the cleaning agent and the solubilized residues from the board and will evaporate readily, leaving it clean and dry. Due to the high cost associated with these solvents, however, their used has been limited to specialized applications. The effectiveness of solvent cleaning on several contaminants is summarized in Table 12.2.

12.2.1 Alcohols

In general, alcohols are defined as a broad class of organic compounds characterized by the formula R–OH, where R is an alkyl radical [59]. From the numerous possible combinations, only a few have found use in electronic production cleaning. These alcohols belong to the lower aliphatic alcohol group, which includes:

   methyl alcohol (methanol, wood alcohol, carbinol);
   ethyl alcohol (ethanol, methyl carbinol, or plain alcohol); and
   isopropyl alcohol (isopropanol, 2-propanol, dimethyl carbinol).

12.2.1.1 Methyl Alcohol

Methyl alcohol, due to its high toxicity, is only used in stable mixtures and azeotropic blends with other less toxic and nonflammable solvents such as CFCs. These so-called lower aliphatic alcohols are characterized by flammability and bipolar solvency☐ that is, they are capable of dissolving polar and nonpolar compounds. Solvency toward polar materials, however, is less than the solvency of water, and solvency toward many nonpolar substances is not as high as exhibited by the nonpolar solvents, such as the chlorinated hydrocarbons used for vapor degreasing [61].

Methyl alcohol can attack the central nervous system of the body whether it is ingested, absorbed through the skin, or inhaled. Very small quantities can cause irreversible brain lesions, which can accumulate to produce such effects as insanity and blindness.

12.2.1.2 Ethyl Alcohol

In its pure form, ethyl alcohol is probably the most effective solvent, but legislation generally insists that it not be made available for industrial purposes without some form of denaturation. Whether an alcohol is usable for an application depends on the denaturing product that is incorporated. If other similar hydrophilic solvents such as pure ketones or heavier alcohols are used for denaturing, ethyl alcohol is usable. On the other hand, if the denaturing product is a petroleum derivative or solid, such as camphor, then foreign matter may recontaminate the board it is supposed to clean [55].

Table 12.2 Relative Solvent Cleaning Effectiveness [74]. With permission from the Institute for Interconnecting and Packaging Electronic Circuits.

Contaminants	Chlorinated	Fluorinated	Azeotrope	Alcohols	Ketones	Aromatics /Aliphatics
Polar						
Fingerprint Salts	I	I	E	E	E	I
Rosin Activators	I	I	E	E	I	I
Activator Residues	I	I	E	E	I	I
Cutting Oils	I	I	G	G	G	I
Temporary Solder Masks	I	I	G	G	I	I
Soldering Salts	I	I	I	I	I	I
Residual Plating	I	I	I	I	I	I
Residual Etching Salts	I	I	I	I	I	I
Neutralizers	I	I	I	I	I	I
Nonpolar						
Resin Waxes	E	E	E	E	E	G
Waxes	E	G	G	I	G	G
Soldering Oils	E	E	E	E	E	E
Cutting Oils	E	E	E	E	E	E
Fingerprint Oils	E	E	E	E	E	E
Flux Resin, Rosin	E	E	E	E	E	G
Markings	E	I	G	I	E	E
Hand Cream	E	E	E	G	E	G
Silicones	E	I	I	I	I	I
Tape Residues	E	E	E	E	E	E
Temporary Solder Masks	E	E	G	I	I	I
Organic Solvent Films	E	E	G	E	E	E
Particulates						
Resin and Fiberglass Debris	M	M	M	M	M	M
Metal and Plastic Machining Debris	M	M	M	M	M	M
Dust	M	M	M	M	M	M
Handling Soils	M	M	M	M	M	M
Lint	M	M	M	M	M	M
Insulation Dust	M	M	M	M	M	M
Hair/Skin	M	M	M	M	M	M

E=Effective, I=Ineffective, G=Moderately effective, M=Mechanical action required

Ethyl alcohol is very hygroscopic and will readily absorb considerable water from atmospheric humidity. If the water content of the alcohol exceeds approximately 1 percent, the effectiveness of the alcohol in removing resin or rosin residues will be diminished; the latter will tend to hydrolyze in the dissolved water, and in this form is extremely difficult to remove by any means.

The principal hazards with the use of alcohol are fire risk and toxicity. The toxicity of pure ethyl alcohol is not high, but can be absorbed into the system through the skin, lungs, or through ingestion, in all cases causing the blood alcohol content to rise. However, the denaturing compounds are frequently more toxic than the alcohol itself, and can be present in the vapors. It is therefore prudent to assume that the vapors of the cleaning alcohol are much more toxic than for pure ethyl alcohol [55].

Because of the fire and toxicity risks, ethyl alcohol, along with other alcohols, is generally used only for cleaning that uses successive cleaner baths in rotation, possibly with assistance from a brush. However, alcohols are frequently added to other solvent types to help them more completely remove flux residues. Whereas most of a rosin compound is soluble in alcohol, some additives and activators, particularly after the thermal reactions at soldering temperatures, will not be soluble. Contamination tests should be conducted after cleaning in order to determine whether the degree of cleanliness is as required [55].

12.2.1.3 Isopropyl Alcohol

Isopropyl alcohol, a by-product of the distillation of petroleum, is cheap, plentiful, of limited toxicity, and does not require denaturing. Thus it is considered one of the ideal alcohols for use in the electronics industry. Several grades are available on the market [40].

The toxicity is higher than for ethyl alcohol but is still sufficiently low for practical flux removal and other applications. The physiological effects due to high absorption include headaches or mild nausea. However, there is no accumulation of toxic substances in the body and after a few hours, the absorbed poison is eliminated. The remarks regarding the hazards of ethyl alcohol apply equally to isopropyl alcohol [40].

12.2.2 Ketones

Like alcohols, ketones can be defined by a simplified general chemical formula R-CO-R, where the two Rs are the alkyl radicals and the CO is from the carbonyl group. Again, numerous compounds fit the general formula, and as with the alcohols, only a few of these compounds are uses in electronic production cleaning [40]. These few compounds belong to the aliphatic group:

∞ methyl ethyl ketone (known as MEK; another term is 2-butanone);
∞ dimethyl ketone (known as acetone; sometimes referred to as 2-propanone); and
∞ methyl isobutyl ketone (known as MIBK; sometimes referred to as 4-methyl-2-pentanone).

These ketones have some commonality with alcohols. They are characterized by bipolar action and flammability, but are chemically more aggressive than alcohols. Ketones attack and dissolve many thermoplastic materials, so users must be careful when cleaning with ketones. In general, ketones provide good solvency for rosin, fats, oils, and greases, and can dissolve small amounts of salts. They are, like alcohols, miscible with water. Ketones are sometimes used for bench cleaning that is, touch cleaning. Their high degree of flammability requires strict rules for handling [40].

12.2.3 Hydrocarbons

Both aliphatic and aromatic hydrocarbons are used in specialty cleaning, such as for magnetic parts or removing organic silicones. Overall, its use is very limited because of flammability. Benzene is a cell poison absorbed through the skin; technical-grade toluene contains appreciable amounts of benzene. Xylene is the safest of these solvents. Toluene, xylene, naphtha, and hexane have good solvency on fats, greases, oils, and rosin [41]. Both aliphatic and aromatic hydrocarbons are immiscible with water and have no solvency on water-soluble matter.

Hydrocarbon solvents are available in various grades of purity; caution is advised when technical-grade solvent is used for cleaning. Naphtha is a mixture of various hydrocarbons, often also containing aromatic solvents. It is available in various grades from highly volatile liquids with very low flash points to liquids with high boiling points and flashpoints [41].

12.2.4 Terpenes

Terpenes represent another solvent class for cleaning printed wiring assemblies. Recovered as by-products of the citrus industry, terpenes have excellent solvency on rosin. Limonene, for example, a natural oil component of lemons, oranges, bergamot, caraway, peppermint, and spearmint, has excellent solvency [41].

Although the surface tension of some terpenes is high, they have demonstrated crevice penetration into spaces of less than 1 mil. In operations using solder flux, post-solder cleaning can be accomplished via spray or underbrush treatment wet brushing the solder side of assemblies after wave soldering. The terpene cleaner dissolves and emulsifies the rosin flux residue; subsequent aqueous spray rinsing removes the emulsion efficiently. Hence, the combination of terpene cleaner and water results in effective bipolar soil removal. The terpene cleaner contains sufficient amounts of surfactant to allow emulsification with the subsequent water rinse. The terpene cleaner will hold more dissolved rosin than its own weight and can achieve greater ionic cleanliness than does the aqueous saponifier process. Solution insulation resistance measurements on specimens cleaned with terpene (and not water rinsed) compared favorably with saponifier-cleaned specimens [41].

12.2.5 Halogenated Solvents

Hydrocarbons containing halogens, such as chlorine and fluorine, are referred to as halogenated compounds. Of the many existing halogenated solvents, a few have achieved commercial significance for electronic production cleaning: 1,1,1 trichloroethane, methylene chloride, and alcohol-, methylene chloride-, and acetone-blended CFCs [58].

These solvents easily remove hydrophobic and nonpolar residues, such as fats, greases, oils, rosin, and resin. The costs of organic solvents are high, compared with those of water, and continuous solvent purification is required for both performance and process economy.

Good cleaning cannot be achieved with a contaminated solvent and continuous use of fresh solvent is expensive. The use of solvent vapor for surface cleaning is a practical solution that assures continuous use of organic solvent without contaminating the parts and without great waste. Vapor cleaning or degreasing uses organic solvent that is continuously evaporated and condensed on parts immersed in the vapor blanket. The condensate dissolves surface soils; the contaminated solvent drips back into the boiling sump, again becoming vaporized, leaving the dissolved contaminants in the sump [58].

Although halogenated vapor cleaning agents are not flammable and have low toxicity, they are subject to thermal decomposition at higher temperatures. Open sparks or flames, overheating elements, inhaled vapors (through a burning cigarette, for instance) can decompose these compounds into very toxic products.

12.2.5.1 Chlorinated

Most chlorinated solvents will remove flux residues to some degree, but they should be carefully considered. In general, chlorinated solvents will attack many plastic compounds and their presence, even as vapor, may be detrimental to many components or their markings. It is therefore essential to test for use on each component before accepting it [61].

All chlorinated solvents must be considered highly toxic, and the quantity of vapor permitted in a workshop is constantly being reduced by health authorities. For example, carbon tetrachloride was commonly used in the industry until it was discovered that process operators suffered irreversible liver damage and that death could results from an accumulation of this solvent. Carbon tetrachloride is now forbidden for use in industry except under the most severely controlled conditions [61].

Likewise, trichlorethylene, initially used to replace carbon tetrachloride, has also been abolished and replaced by perchlorethylene. However, it is known that like trichlorethylene, perchlorethylene can also cause permanent hepatic lesions, so the utmost precautions should be taken to protect operators from contact with either the liquid or the vapors [61].

Chlorinated solvents are commonly used for degreasing, often in vapor installations. However, generally speaking, their boiling points are too high to allow vapor installations to be used for cleaning printed circuits. The solvents are therefore generally used as "cold" cleaners□ that is, circuit cards are immersed in either the pure solvents or in mixtures with alcohols and/or ketones. A variation sometimes used is to brush only the solder side of the circuit with these solvents in an automatic conveyorized machine. These machines are reasonably effective for a general cleaning, but are not suitable for high-quality applications. The brushes can get contaminated with dissolved resin residues and recontaminate clean circuits; moreover, the component side is not cleaned at all. Therefore, this type of machine should not be used for high reliability applications, except as a pre-cleaner before some other cleaning method [61].

12.2.5.2 Fluorinated

Fluorinated solvents are very popular for flux removal in the electronics industry because they combine reasonable dissolving of greases with a comparatively low toxicity. However, they do require precautions. The most common fluorinated solvent is a 1-1-2-trichloro-1-1-2-trifluoroethane usually abbreviated F-113 and sold by a number of companies under trade names such as Freon, Arklone, Flugene, and Frigene. Pure F-113 is not a good solvent, with a limited capacity for flux removal. For this reason, the solvents are generally blended with various alcohols, chlorinated solvents, or aromatic solvents. This blending may be done to render the mixture azeotropic□ that is, when it is distilled, the distillate will have the same chemical composition as the raw product. Other blends may be non-azeotropic□ for example if they contain higher quantities of alcohol than of the azeotropic solvent□ and these are generally better flux removers than the azeotropes. However, during distillation nonazeotropic mixtures distill off as an azeotrope, leaving a higher concentration of alcohols in the tank. Introducing a thermostat into the tank, to cut off the heat at, say, 60℃ eliminates this disadvantage, since the circuits are first placed in the boiling tank, where the high alcohol concentration will provide the most effective flux removal. Since the solvent vapor above the boiling tank is azeotropic, containing from 3 to 5 percent alcohol, there is no fire hazard, even though the percentage of alcohol in the boiling tank may be considerably higher. Once the mixture in the tank reaches the thermostat setting, the machine will cut off and the tank, with its accumulation of rosin, can be emptied for disposal and refilled with fresh solvent [55].

At least thirty or forty solvent blends are available on the market from different suppliers. Due to toxicity, no blend containing methyl alcohol should be used.

Blends containing methylene chloride should be tested with caution to ensure they do not attack either plastic parts or the epoxy resin substrate. The efficiency of these solvents is relatively low, even if used in vapor phase, but their advantages may outweigh their disadvantages. A selected combination of flux and solvent must be tested at regular intervals for ionic contamination after cleaning. Many activators are not soluble in these solvents and nonvisible highly ionic activator residues may be left on the circuit. This can occur with any type of activator, but may vary considerably according to minor parameter changes in the soldering process. Thus, it is not sufficient to perform a single test; tests must be done on a regular basis [55].

12.2.5.3 Decomposition

Chlorinated and fluorinated solvents can decompose in two ways, one of which can be detrimental to printed circuits, and the other detrimental to process operators.

The first is hydrolytic decomposition of the solvent. In the presence of acid (such as rosin) and moisture (atmospheric humidity) pure chlorinated solvents will decompose, forming hydrochloric acid as one by-product. Hydrochloric acid on a printed circuit can cause corrosion, leakage currents, and many other detrimental effects. The acidity of dissolved fluxes is sufficient to start this decomposition, which can be particularly bad if two dissimilar metals are in contact, which is common in the typical printed circuit. To avoid this decomposition, the majority of commercial chlorinated solvents are stabilized with an amine product, which is itself a form of ionic contamination. If these solvents are used and particularly if they are redistilled, they must be regularly monitored to ensure that sufficient stabilizer is present. This is done with an acid acceptation test; the supplier of the solvent should be consulted as to how to proceed. Acid acceptation tests should be conducted at least on a daily basis. As the acid acceptation approaches zero, the solvents should be destroyed rather than sent to a recuperator [61].

The other decomposition phenomenon associated with chlorinated and fluorinated solvents is the formation of toxic gases as vapors in contact with flames, incandescent objects, or catalytic surfaces. In these cases, the solvent breaks down into phosgene, which is a highly toxic gas causing paralysis and death. No object whose temperature exceeds 250℃ should be permitted in the same room as chlorinated solvents. Of course, this includes an absolute interdiction on smoking, as the respiration of even a small quantity of chlorinated solvent vapors through an incandescent cigarette will create sufficient phosgene to cause problems. One of the catalytic compounds that will provoke the formation of phosgene is tin oxide at temperatures exceeding about 150℃. Since inevitably hot tin oxide is associated with every soldering machine or soldering iron, none of these solvents should be used in the same room as any soldering operation.

Exhaust ventilation should be done at floor level of every room where halogenated solvents are used. These solvents are inevitably heavier than air and if vapors escape from a process, they tend to accumulate at floor level. It is not unknown for operators stooping down to pick something up to be suffocated by the accumulation of solvent vapors floor level due to lack of oxygen. Gas masks should be worn when cleaning out machines or containers that have contained halogenated solvents; this should always be done under supervision.

12.2.6 Solvent Cleaning Equipment

Batch solvent cleaners are mainly used for low-volume cleaning applications (Figure 12.1). The main elements of most batch cleaners are the main tank (with one or two sumps), the condensing coils, the spray wand, and the hoist. A system with two sumps, one boil sump, and one rinse sump is preferred. Clean solvent from the condensing coils is returned to the rinse sump, which overflows into the boil sump. The boil sump generates the solvent vapor [56].

The board assembly is lowered into the vapor blanket to begin the cleaning cycle. The vapor condenses on the assembly, which begins the process of contamination removal. The board should be placed in the vapor for at least two minutes. If additional cleaning is required, the board may also be sprayed with the wand for about a minute below the vapor line, or immersed in the rinse sump for about a minute. To minimize vapor loss, the board should be removed from the vapor slowly (preferably with an automatic hoist), and a cover should be placed over the main tank when the cleaner is not in use [56].

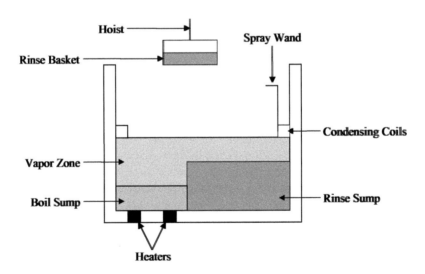

Figure 12.1 Batch solvent cleaning system. With permission from CALCE Electronic Products and Systems Center.

In-line solvent cleaners are intended for medium- to high-volume applications, or situations where better control of the cleaning process is required. These cleaners operate much like batch solvent cleaners, but the process is much more automated and more stable since operator-induced variables are reduced.

A typical in-line solvent cleaner (Figure 12.2) consists of five different zones: entrance vapor zone, pre-clean vapor zone, pre-clean spray zone, spray/immersion zone, and exit vapor zone. There are numerous variations on this basic configuration. Four sumps are usually used: a boil sump, a pre-clean sump, a spray/immersion sump, and a distillate sump. The solvent cascades from the distillate sump to the boil sump. Condensing coils are used to contain the vapor. The board assembly is transported through the cleaner on a stainless steel mesh belt, which is inclined on both ends and horizontal in the center. Plastic or metal cleats are commonly attached to the mesh belt to prevent the board from sliding down the incline. The spray nozzle angle, pattern, and pressure can be varied in most systems. The top spray pressure should be set higher than the bottom pressure to prevent the board from being lifted off the mesh [56].

The mesh belt slowly transfers the board assembly into the entrance vapor zone, where the solvent vapor condenses on the board, starting the cleaning process. The board is sprayed top and bottom in the pre-clean spray zone and completely covered with solvent in the spray/immersion zone; the overhead spray agitates the solvent. As a final rinse, the distillate spray zone sprays the top and bottom of the board with very clean solvent. The board then passes through the exit vapor zone and out of the cleaner [56].

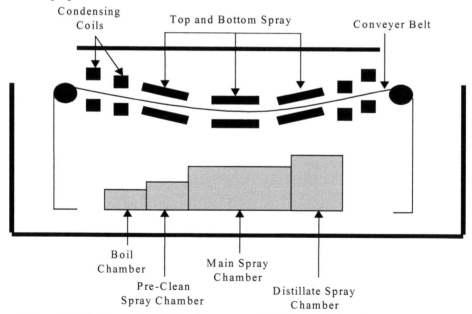

Figure 12.2 In-line solvent cleaning system. With permission from CALCE Electronic Products and Systems Center.

12.2.7 Safety and Environmental Concerns

All process media, when used by poorly trained personnel or carelessly, can present hazards to operations personnel. Physical and chemical hazards must be prevented by training personnel, providing effective ventilation, and using the appropriate equipment.

All the cleaning agents discussed so far can pose a threat to the environment unless they are used with adequately designed equipment, good recovery and treatment systems, good controls, and by well-trained personnel. All personnel involved should consider emissions into the atmosphere, bodies of water or water treatment systems, and conservation of energy and water resources.

The greenhouse effect, ozone layer depletion, acid rain, groundwater pollution, uncontrolled toxic waste dumping, depletion of fossil fuel, and water shortages are all problems that need to be addressed, even though appreciable capital expenditures may be required.

12.2.8 Solvent Alternatives

Because of the problem of ozone depletion and the Montreal Protocol, new materials are being actively sought to replace old solvents. The search for new materials has been especially intense in the case of a substitute for CFC-113 [4].

The new solvents are generally known as stratospherically safe fluorocarbons (SSFs). The principal solvent molecules to emerge as SSFs are the hydrochlorofluorocarbons (HCFCs), which contain hydrogen, chlorine, fluorine, and carbon. For the electronics industry, the two most important HCFCs are HCFC-141b and HCFC-123. HCFCs can damage acrylic based polymers by crazing, cracking, and delamination.

12.3 Aqueous Cleaning

As a well-known technology with a long history, aqueous cleaning uses water as a solvent to remove residues, but not alone. Generally, water is teamed with a 5 to 15 percent concentration of saponifier-alkalis in which a chemical reaction forms with acids in insoluble rosin-based flux to form a soluble rosin "soap;" this is then softened and removed via a deionized water rinse.

Water-soluble fluxes can be cleaned with water alone. However, they have very low surface tension, which permits them to flow under components with low standoffs; water, with a rather high surface tension, is sometimes prevented from reaching these corrosive materials. Accordingly, saponifiers are used to reduce water's surface tension and promote its flow into tight spaces. Again, washing must be followed with a deionized water rinse, also requiring drying [75].

To fill specific functions in contamination removal, aqueous solutions contain additives, such as acidic reacting solutions, basic reacting solutions, complexing agents, foam control agents, or wetting agents. Any operation with these additives must be followed by pure water rinsing to prevent the formation of additional contamination from the PWB assembly materials. The relative effectiveness of aqueous detergent cleaning is summarized in Table 12.3 [75].

12.3.1 Water

The use of water and water cleaning systems has been popular since the early 1970s. Water is a solvent of very high polarity. Its nearly universal availability, safety, and excellent ability to dissolve ionic materials make it an attractive choice for cleaning. However, moisture that may have been absorbed in porous insulating surfaces or capillary spaces in or under components may have to be removed. Moisture retained after cleaning can result in electrical degradation or failure due to the intrinsic conductivity caused by any ionic materials dissolving in the water.

Water, alone or with additives such as rinse aids or neutralizers, can be used to remove the residues of water-soluble fluxes. Water with saponifiers can be used to remove rosin-base fluxes.

12.3.2 Neutralizers and Rinse Aids

Water-soluble fluxes are, as their name implies, soluble in water; however, their exposure to the heating of the soldering process often partially converts them to reaction products that are less soluble than the initial flux. This can be caused by oxidation, pyrolysis, or other chemical reactions. The high-pressure flushing action of an aqueous cleaning system can often partially compensate for lack of residue solubility by dispersing insoluble materials.

High levels of ionic residues can occur in the finished assembly even though it appears visually clean. Additives, such as neutralizers or rinse aids, in the wash water can significantly improve ionic cleanliness and, for many types of residues, the visual appearance of the board. These materials are typically alkaline solutions containing ammonia, amines, or other alkaline compounds, along with surfactants, defoaming materials and, perhaps, chelating agents. Although additives assist in dissolving flux residues, they too must be thoroughly rinsed away because their residues can be ionic and conductive [40].

Table 12.3 Relative Aqueous Detergent Effectiveness [75]. With permission from the Institute for Interconnecting and Packaging Electronic Circuits.

Contaminants	Effective	Moderately Effective	Ineffective
Polar			
Fingerprint Salts	X		
Rosin Activators	X		
Activator Residues	X		
Cutting Oils		X	
Temporary Solder Masks	X		
Soldering Salts	X		
Residual Plating	X		
Residual Etching Salts	X		
Neutralizers	X		
Ethanolamine	X		
Nonpolar			
Resin Waxes	X		
Waxes		X	
Soldering Oils	X		
Cutting Oils	X		
Fingerprint Oils	X		
Flux Resin, Rosin	X		
Markings		X	
Hand Cream		X	
Silicones			X
Tape Residues			X
Temporary Solder Masks	X		
Organic Solvent Films	X		
Surfactants	X		
Particulates			
Resin and Fiberglass Debris			X
Metal and Plastic Machining Debris			X
Dust	X		
Handling Soils		X	
Lint		X	
Insulation Dust		X	
Hair/Skin		X	

12.3.3 Saponifiers

Although rosin is not water-soluble, it is possible to react the rosin in a highly alkaline solution to form a rosin soap that is very water-soluble. This process is called saponification. The major active ingredients of saponifier chemicals are alkaline materials like amines. In aqueous solution, these materials react with rosin acids, the major constituents of gum rosin, in a simple acid-base reaction.

While any strong alkaline material can be used to dissolve rosin in a similar way, solutions of selected alkanol amines are most frequently used. These solutions provide a relatively limited pH range of approximately 10.5 to 11.8. Higher pH values may attack component markings, coatings, or the plastic laminates themselves [41]. Of particular concern are metallic inks often used on integrated circuit components for identification. The inks are usually pigmented with finely divided aluminum flakes, which in the presence of highly alkaline solutions can react and dissolve. This type of marking material should therefore be avoided on any components that will be cleaned in saponifiers.

At pH values below the 10.5 to 11.8 range, saponifiers remove rosin at considerably slower rates□ possibly incomplete in the washing time allowed. Therefore, the alkaline materials used in saponifiers must provide a buffering action in the desired range.

Saponifiers are generally supplied by the manufacturer as a concentrate that must be diluted in water before using. Because the content of the active ingredient (amine) varies widely from one product to another, the amount of concentrate required will differ. Careful attention should be given to manufacturer recommendations for bath composition.

Although the predominant composition of natural gum rosin is saponifiable acids, as much as 10 percent or more can be unsaponifiable material that will neither react nor be dissolved by an alkaline solution. To aid in removing this residue, higher-boiling solvents, such as glycol ethers, are included in the saponifier. These are water-soluble and capable of softening or dissolving the unsaponified portions of the residue for easier removal. Various surfactants are also added to help remove and disperse any residual particulate matter. These agents, together with the vigorous flushing action of high-pressure, heated sprays, serve to remove any unsaponifiable residues.

During use, considerable amounts of rosin can be reacted and dissolved in the heated saponifier solution. The major by-products of this reaction are rosin soaps. Like other water-soluble soaps, they have a tendency to generate foam, particularly with the high-pressure spray and agitation encountered in commercial aqueous cleaning equipment. As the rosin soap content of the wash solution gradually rises, the amount of foam likewise increases. In time, excessive build-up of foam in the system may clog spray jets or cause overflow, compromising the cleaning efficiency of the entire system. Changing the wash solution or adding defoamers helps avoid this problem [41].

12.3.4 Defoamers

Defoamers are usually initial constituents of the saponifier, but additional defoamers can be added to help control foam and prolong the useful life of the saponifier wash solution. In most instances, the life of a saponifier solution is limited not by the depletion of the active alkaline material, but by the gradual build-up of rosin soaps and foam.

In the original saponifier mixture, low foaming, nonionic surfactants are generally used. The defoaming properties of these materials are very temperature-dependent. Such surfactants generally show increasing defoaming properties as the temperature is increased. If operating temperatures are excessive, however, the useful life of the bath can be limited by the volatilization of some of the active ingredients in solution. Because a variety of foam-control materials are used by different vendors in different products, it is important to follow the guidelines of the manufacturer for the operational temperatures of each product [55].

In some instances, depending on the agitation provided by the equipment or the volume of the product being processed in the bath, foaming cannot be controlled even though the system is operating within the recommended temperature range. Under these circumstances, the gradual build-up of foam can be controlled and the bath life extended by adding more defoaming materials. A variety of types, including silicone and nonsilicone defoamers, can be added. If bonding or coating is to follow the cleaning, the defoaming materials must not interfere with adhesion.

12.3.5 Surfactants

Surfactants are used either as integral ingredients of a saponifier or detergent or alone in the wash water. "Surfactant" is a contraction derived from *surf*ace *act*ive *agent*. Surfactants are capable of lowering the surface tension of the solvent (usually water) to make the solution wet better and penetrate capillary spaces, crevices, and pores. This improved wetting helps the water flush away trapped residues more thoroughly [41].

Surfactants are commonly classified by the ionic character of the major active part of the molecule (nonionic, anionic, or cationic). Because the nonionic surfactant molecules carry no electrical charge, their performance is not affected by other ionic materials in solution, such as those found in hard water.

In addition to their use in aqueous cleaning formulations, nonionic surfactants are widely used in water-soluble soldering fluxes and heat-transfer fluids, such as soldering oils. Many of these surfactants show considerable thermal stability, even at temperatures up to 250 and 300℃, for limited exposure times. This heat stability, together with a generally excellent rinsability in water, makes them preferred choices for use in both soldering and cleaning operations [41].

12.3.6 Aqueous Cleaning Equipment

Batch aqueous cleaners are used principally for low-volume cleaning applications. They contain a main tank, a basket for holding the board assembly during cleaning, an overhead spray system, and a convection dryer. A programmable controller establishes pre-rinse, wash, post-rinse, and drying times, and the number of wash cycles. If a saponifier is used, the cleaner will add and monitor the chemicals. One advantage of this type of system is that the boards are vertical, which allows the water and contaminants to efficiently drain. Since the volume of water on the board surface is kept to a minimum, drying is also easier [56].

In the cleaning cycle the assemblies are placed in the basket, the door is closed, and the controller is programmed with the proper information. A quick pre-rinse removes most of the contamination, and is followed by one or more extended wash cycles. After the post-rinse, the convection dryer circulates hot, dry air through the cleaner to dry the assemblies.

In-line aqueous cleaners are intended for medium- to high-volume applications. Unlike solvent cleaners, a batch aqueous cleaner may actually provide better process control than an in-line aqueous cleaner. A typical in-line aqueous cleaner consists of at least five different modules: pre-rinse with air knife, wash #1 with air knife, wash #2 with air knife, post-rinse with air knife, and final dry. The spray nozzle angle, pattern, and pressure may be varied, but as in the solvent in-line system, the top spray pressure is higher than the underside pressure to keep the assemblies on the mesh belt.

Placing the assemblies on the mesh belt, which slowly transfers them to the cleaner, starts the process. The assemblies are sprayed top and bottom in each module. Excess water is removed between modules with an air knife. Final drying occurs in the dryer module. Since the assemblies are horizontal, water can puddle on the surface, making drying more difficult.

12.4 Semiaqueous Cleaning

With semiaqueous cleaning, an organic solvent is used to dissolve manufacturing residues, leaving a residue that is easily soluble in water. In other words, the semiaqueous process uses solvent to remove manufacturing residues, and water to remove solvent residues. As in the aqueous process, a final rinse with deionized water should be used [57].

The most common semiaqueous solvents used are terpenes, aliphatic hydrocarbons, diglycol ethers, and alkoxypropanols. Suppliers claim that these solvents are generally biodegradable, nontoxic, and noncorrosive. Since they contain no chlorine, their ozone depletion potential is zero. However, most of them are classified as volatile organic compounds (VOCs).

Batch semiaqueous cleaners are used primarily for low-volume cleaning applications. Often two units are used with this system: one for the solvent wash, and the other for the aqueous rinse. Both units contain a main tank, a basket for holding the assemblies during cleaning, and an overhead spray system; the aqueous unit also contains a convection dryer. Semiaqueous solvents are flammable, so some method of fire control, such as an inert atmosphere, is generally required. Both units use a programmable controller to establish pre-rinse times, number of wash cycles, and drying time. Both units hold the assemblies vertically so contaminants, solvent, and water drain off, making drying easier [57].

The cleaning cycle begins by placing the assemblies in the solvent wash unit and programming the controller. A short pre-rinse is followed by one or more extended wash cycles and a post-rinse. The assemblies are then placed in the aqueous unit and the controller is programmed to cycle through a series of pre-rinses, washes, and post-rinses to remove the solvent. Circulating hot, dry air through the cleaner dries the assemblies.

In-line semiaqueous cleaners are intended for medium- to high-volume applications. As with aqueous cleaners, the batch semiaqueous cleaner may offer better process control than an in-line semiaqueous cleaner. In-line semiaqueous cleaners are actually two cleaners in one. The first half of the cleaner is devoted to washing the assemblies with semiaqueous solvent, and the second half uses water to remove the solvent. A major concern here is the solvent mist does not ignite and cause a fire. One method of fire control is to use an inert atmosphere inside the solvent module; another method is to use a spray-under-immersion system to prevent the mist from occurring. An air knife removes as much solvent as possible before the assembly moves into the aqueous wash. As with in-line aqueous cleaners, a series of modules consisting of spray nozzles and air knives is used to wash the assembly with water. Vapor loss is not a concern, so a horizontal stainless steel mesh belt can be used. The spray nozzle angle, pattern, and pressure can be varied in most systems, with the top nozzle pressure, again, higher than the bottom side pressure.

The assembly is placed on the mesh belt and transferred into the solvent module, where the assembly is sprayed top and bottom with solvent. An air knife is used to remove excess solvent before transferring the assembly to one or more aqueous modules, which spray the top and bottom of the assemblies with water; the excess water is removed with an air knife. Final drying occurs in the drying module with the assembly horizontal, which makes drying more difficult [57].

12.5 Emulsion Cleaning

Emulsion cleaning is a combination of aqueous and semiaqueous cleaning which uses any two immiscible (nonmixable) liquids, a mixture of nonpolar organic solvents and nonionic surfactants. Soldered assemblies are first washed in the concentrated blend of the cleaner. The solvent portion of the cleaner dissolves any rosin, oils, or other low-polarity residues left after soldering. The assemblies with the residual solvent and surfactant can then be put through a conventional aqueous cleaning process. Although the solvent portion of the cleaner is not water-soluble, the surfactant allows the cleaner to be emulsified and dispersed in the water as discrete microscopic droplets. Subsequent rinse stages can further flush away any traces of the cleaner [19].

Thus far, the process has been developed using terpene solvents one example is d-limonene, which is found in the oils extracted from oranges. This same principle can be applied to other suitable solvents and surfactant mixtures.

Processes using these materials have some benefits in terms of both solvent and aqueous cleaning. They are really two processes using a solvent and an aqueous cleaning system. In the solvent cleaning section, nonpolar organic residues such as rosin are efficiently dissolved by an effective nonpolar solvent, such as a terpene. In the subsequent aqueous cleaning section, any residual polar or ionic materials are efficiently removed, along with any solvent residues, by water washing and rinsing. Surfactants, along with the terpene, enable its complete removal by emulsification in water. Additionally, the surfactants will aid in wetting and dispersing any non-soluble particles that may be in the residue. The dual solvent cleaning process allows the use of optimal cleaners for each of the constituents in the flux residue and can result in cleaner assemblies with lower levels of ionic residues [19].

12.6 Ultrasonic Cleaning

Ultrasonics can be used to clean or to improve a less aggressive solvent's cleaning ability. Ultrasonic transducers are mounted on the exterior of the cleaning tank to vibrate its walls; this agitation creates microscopic air bubbles in the solvent that expand and contract, giving extra scrubbing power [19].

Historically, the military has been against the use of ultrasonic energy for cleaning, because older studies suggested that the high-frequency vibrations were generating fatigue and breakage of the interconnects inside some components. The military has since changed its view, because newer wire bonding techniques have provided more robustly interconnected components able to withstand ultrasonic vibrations, and because the military began using components in which the wire bonds were encapsulated. This is not to say the process is 100 percent safe. Under some conditions, ultrasonic energy can fatigue some electronic elements [19].

12.7 Plasma Cleaning

Plasma cleaning removes organic contaminants by creating plasma from an appropriate gas, usually oxygen, and submersing the assembly into the plasma. Plasma cleaning is more environmentally friendly than solvent cleaning by generating less waste [147].

Plasma cleaning has limitations. Oxygen plasma, for example, can cause silver surfaces to oxidize and thus requires argon plasma to remove the silver oxide. Oxidation of other metals, such as those used for resistors, can cause undesirable changes in resistivity. The removal of aluminum and its sputtering over other surfaces can be detrimental to some circuitry.

Many of the problems with gas plasma cleaning are time-related; premature termination of the plasma action will result in incomplete decontamination or stripping. Therefore, it is important to determine the optimum time of exposure for acceptable cleaning or stripping [49].

12.8 Selecting a Cleaning Method

Selecting a cleaning chemistry that is right depends on several variables. Deciding which media work best for cleaning is only part of the picture. Materials compatibility must determine how cleaning solvents will react to other materials on the board, such as component markings, soldermasks, laminates, adhesives, and polymers. Industrial hygiene and safety addresses solvent odor, flammability, and exposure limits. From an environmental standpoint, it must be determined whether to treat waste as a hazardous material, a recyclable, or safe drainage. Similarly, alternative cleaners should be identified if potentially "risky" cleaners are being used.

The cost of a cleaning process must be considered, including equipment, materials, maintenance, engineering, labor, training, and operation. There are also costs associated with support equipment for water, air, and waste management. The chemistry and the equipment may have to be chosen together. Batch versus in-line, wash time and temperature, rinse time and temperature, spray pressures and angles, conveyor speed, water feed quality, drying time and temperature, and the use of ultrasonic energy are vital chemistry related parameters.

With the increased use of water for cleaning, preparing the wastewater for discharge into a public treatment system, called pretreatment, is a growing concern. Pretreatment factors include water temperature, pH level, heavy metal content, chemical oxygen demand, biological oxygen demand, total toxic organics, and other contaminants, such as oil and grease. Specific requirements vary from location to location. Pretreatment equipment may have to be purchased to provide proper waste management. Closed loop aqueous recycling systems are available to minimize the discharge of water.

Cleaning media and application techniques, if not carefully chosen, controlled, and used, may be the cause and origin of many different types of contamination. Ironically, cleaning agents and processes, while removing some types of contaminants, can introduce others at the same time. The basic causes of contamination via cleaning are incompatible cleaning media, cleaning media contaminated during use, and impure cleaning media. When a cleaning agent is capable of dissolving a material in the assembly, the dissolved substance can be spread by the solvent onto other parts of the assembly. This type of contamination is most prevalent in, but not confined to, immersion cleaning. The degree of contamination will depend on solvency, production volume, shop practices, and the number of stations used.

Postsolder cleaning, often called defluxing, of a printed circuit card assembly is often considered a cost adder rather than a value enhancer. However, freedom from harmful contaminants is as important to reliability over the product life cycle as the robustness of the soldered interconnections. Ionic contamination in humid environments can lead to corrosion, electromigration, and failure of conformal coatings to adhere to the substrate. This is not to say that one cannot eliminate postsolder deflux by the use of synthetic based, low-solids fluxes and a control fabrication process. The viability of this approach depends on solderability issues, electrical testing impacts, properties of the flux residues, product operating environments, and life-cycle requirements. Design characteristics that increase the possibility of entrapment and the difficulty of removing aggressive flux residues also increase concern about the quantity and effect of the low-solids/no-clean fluxes left in place.

12.9 Drying

Water and other volatile liquids absorbed or present as surface films often need to be removed. The rate of evaporation of these materials at room temperature varies with their boiling points and vapor pressure and with the ambient conditions. Accelerated removal techniques include heating, vacuum drying, forced blow-off, and displacement by other liquids.

The tolerable amount of volatiles remaining on or in the parts varies with the type of subsequent process steps, the applicable quality standards, and electrical performance requirements. In the electronic assembly production process, drying may occur prior to:

- ∞ resist application in substrate fabrication;
- ∞ multilayer board lamination;
- ∞ solder mask application;
- ∞ substrate packing and shipping;
- ∞ soldering; and
- ∞ coating, sealing, bonding, potting, and molding.

To determine how dry the assembly should be each type of part should be checked to establish the optimal method, conditions, and duration of the intended treatment. Parts of like construction and geometry can be treated similarly. Care should be taken when establishing profiles for complicated assembly configurations, since some areas might not be exposed to the treatment as well as others [4]. A common test practice for establishing a practical end point for drying is to weigh small assemblies or sections before and after drying stages to determine the best drying profile to reach a constant weight.

Heat and vacuum are two methods for drying. However, neither can conventionally remove bonded water, unless the temperature used is high enough to break the bond.

12.9.1 Heat

Heat, the most common method of removing volatile matter, is applied to both continuous conveyorized and batch processing operations after exposure to aqueous media. Only short heating times are practically achievable with conveyorized operations, but drying can be improved by using a preceding air-knife treatment that eliminates most of the surface water. The dryness achievable this way is adequate for assemblies that do not need manual soldering, bonding, and coating.

Ovens used for batch drying are set for a specific temperature with a typical variation of ±5℃ from the set point. The ovens used for this purpose should have controlled temperatures within this range, fast temperature recovery for removing and loading batches, and adequate heat capacity for the thermal mass to be produced. All ovens should also follow provisions for intake air cleanliness.

Heat drying is generally a long process, if it is necessary to remove more than 90 percent of the moisture, because there are usually limits on the maximum temperature allowed by the electronics being dried. Other methods will remove more moisture quicker and at a lower temperature.

12.9.2 Vacuum

The boiling point of a liquid is the temperature at which the vapor pressure of the liquid exceeds the atmospheric pressure. The vapor pressure of liquids will cause their eventual evaporation at room temperatures, such as 23℃ and 760 mm Hg. The rate of evaporation depends on the vapor pressure of the liquid. A liquid with a relatively high vapor pressure, such as methylene chloride (about 440 mm Hg) will evaporate very fast (compared with water, with a vapor pressure of 21 mm Hg). At the boiling point, evaporation is fast. When the pressure is reduced, so is the boiling point, and lower temperatures are then capable of achieving the same rate of evaporation [61].

12.10   No Clean

It can be argued that not all electronic assemblies may need to be cleaned. Many facilities are opting for the "no clean" process, but the "no clean" process will only work if the boards and components are clean prior to starting this process. Although its goal is to save the expense of cleaning by eliminating the procedure entirely, the strategy is to manufacture a "soil-free" product that *needs* no cleaning. The first step in the process is to select a flux and/or solder paste with activators that will not cause reliability problems as a residue.

However, residues from no clean fluxes may cause contamination problems. Several studies conducted at the National Center for Manufacturing Sciences show that while ionics tend not to be a problem with no clean fluxes, weak organic acids in no clean fluxes may cause current leakage [148]. For this reason, no clean fluxes should be applied with control so that the entire assembly is not sprayed with flux. They should not be regarded as completely benign substances that cannot cause reliability problems.

PART 4

CONFORMAL COATINGS

CHAPTER 13   CONFORMAL COATINGS

There are two categories of circuit board coatings: bare-board coatings and assembly coatings [25]. Bare-board coatings, also called permanent solder masks, are applied during the board fabrication process as liquids or dry films over conductive finishes or bare copper, and are cured by either heat or ultraviolet (UV) light. The primary function of the solder mask is to ward off solder wetting, reducing the need for touch-up after machine soldering.

Assembly coatings, also called conformal coatings, are thin layers of synthetic resins or plastics that are normally applied to electronic assemblies. They are "conformal" in that they conform to the contours of the assembly.   Their purpose is to protect the electronic assembly against a variety of environmental, mechanical, electrical, and chemical problems including contaminants such as dust, dirt, fungus, moisture, chemicals encountered during field conditions. Conformal coatings may also protect the circuit card assembly from alpha particles, corrosion, thermomechanical stress, mechanical shock, and vibration. They are "conformal" in that they conform to the contours of the assembly.

The use of conformal coatings for printed circuits may enable the design engineer to achieve narrower conductor lines and closer spacing. If coatings are not used, impurities, moisture, and other contaminants can bridge the conductors, causing decreases in the insulation resistance or arcing between conductors [107].

The development of conformal coating technology was fostered largely by the requirements of military and aerospace industries. However, conformal coatings have also been used in selected telecommunications, industrial controls and instrumentation, consumer electronic, and automotive applications. In fact, between 1989 and 1998, the use of conformal coatings in North America has grown at an average annual growth rate of five percent even though defense spending has dropped considerably. Overseas usage increased at an average rate of twenty-four percent per annum [37].

13.1 Types of Conformal Coatings

Conformal coatings are generally classified according to the molecular structure of their polymer backbone. Five major types of conformal coatings are used in the electronics industry: urethanes, silicones, acrylics, epoxies, and parylenes. The most widely used conformal coatings are covered by U.S. military specification MIL-I-46058, which defines the following types of conformal coatings: urethane (polyurethane), silicone, acrylic, epoxy, parylene (polyxylylene).

Urethane (polyurethane) and epoxy are general-purpose coatings. Silicone is used for applications requiring resistance to high temperatures, and parylene (polyxylyene) is used when a uniform, pinhole-free coating is necessary [107].

Parylene is the only coating applied via vacuum deposition. The other conformal coatings can be applied by manual brushing, dipping in a reservoir, or spraying, either manually or by machine.

13.1.1 Urethanes (UR)

Urethanes are highly popular in the electronics industry because they have good mechanical properties, good adhesion, and high chemical and moisture resistance at lower costs. They are available as solvent-based (two component formulation), single-component formulations and can be either moisture- or air-cured to yield a thermoplastic material. They can also be obtained in a volatile organic compound (VOC)-exempt composition [26].

Single component urethane requires three to ten days at room temperature for final cure [106]. In the two-part systems, cleaning and disposal are problematic, but curing takes only a few hours, rather than days. These systems result in a thermoset material when cured.

Urethanes have excellent dielectric properties, are resistant to abrasion, and are curable by ultraviolet light. Rework is relatively difficult, but can be accomplished by stripping or by heating the coating material to decompose it into carbon dioxide and water. However, stripping of the coating can cause corrosive damage to the assembly.

13.1.2 Silicone (SR)

Silicone is often used in high-temperature environments because it is soft and flexible. However, silicone is a relatively expensive coating. It is available in three forms: a solvent-based, one-part compound that is cured either at room temperature or by applying heat; a solventless, one-part compound that is moisture-cured; and a two-part compound that can be cured either at room temperature or by applying heat [35].

Silicone is usually applied in thicker quantities than other coatings, thereby providing good resistance to abrasion. Rework is relatively difficult, but is possible by scraping, blasting, or stripping the surface.

13.1.3 Acrylic (AR)

Acrylic is the easiest conformal coating to apply but is not very resistant to chemicals and abrasion. It is usually an organic-solvent-based compound, but can also be obtained in volatile organic compounds (VOC)-exempt form. Typically it is applied as a single-component system, resulting in a thermoplastic coating. The two-component acrylics cure upon mixing to give a thermoset material applied by brushing, dipping, or spraying. Repairs are easily accommodated through the use of vapor degreasers, VOC-exempt solvents, blasting, or alcohol. The traditional solvent-based materials are easily removed; so, they do not provide good protection against solvents [35]. However acrylics have the highest humidity resistance of any of the available range of resins [106].

13.1.4 Epoxy (ER)

Epoxy has excellent abrasive resistance and reasonable humidity resistance. They are fairly easy to apply, but very difficult to remove. They are very difficult to remove chemically for rework since any stripper that will remove the coating will vigorously attack the epoxy potted components as well as the epoxy glass board itself. The only effective way to remove the epoxy coating is to burn through the epoxy coating with a knife or soldering iron.

Epoxy systems are usually available as two part, amine-reacted compound, which cures into a thermoset coating [107]. The coating helps increase system integrity for assemblies that are subject to vibration loading, although unwanted mechanical stresses may be created during curing in some sensitive components, such as glass diodes, due to shrinkage. Therefore when epoxies are applied, a 'buffer' material must be used around fragile components to prevent their damage from film shrinkage during polymerization [106].

Repair, though difficult, is performed by burning or stripping with a solvent. If care is not taken, excessive heat from burning can damage the components, and the active agents in the solvents can attack the printed wiring board laminates and component packages [35].

13.1.5 Parylene (XY)

Parylene, which is inherently solvent-free, is vacuum-deposited and offers resistance to humidity, moisture, abrasion, high temperatures, and chemicals. The three basic forms of parylene are classified as types N, D, and C. Type N is noted for its strong penetrating power, and type D for its thermal stability. Type C parylene offers the most broadly useful combination of electrical and barrier properties, and is particularly well suited to fluorescent activation, a process by which the parylene is made sensitive to UV light. This is accomplished by exposing the coating to radio frequency (RF)-excited gas in a plasma chamber, a process that does not change the material properties of the parylene. The most effective gas for exciting the parylene is argon; however, other gases, such as nitrogen and helium, may also be used. Fluorescent activation aids the visual inspection process [39].

Parylene N: Parylene N is a primary dielectric, exhibiting a very low dissipation factor, high dielectric strength, and a dielectric constant invariant with frequency. This form of parylene has the highest penetrating power [102]. The lighter Parylene N monomer molecule with no chlorine atom in its molecule resists condensation and polymerization and stays in the gaseous state for a longer period compared to the other two parylenes, allowing for greater penetration of deep recesses before the polymer film is formed [109].

Parylene C: Parylene C has a useful combination of electrical and physical properties plus a very low permeability to moisture (due to a chlorine atom on the benzene ring in its molecule) and other corrosive gases. Along with its ability to provide a true pinhole free conformal insulation, Parylene C is the material of choice for coating critical electronic assemblies [102].

Parylene D: Parylene D, with two chlorine atoms on the benzene ring in its molecule, has the highest thermal stability of the three primary parylenes, offering superior physical and electrical properties at high temperature [109]. However, the heavier Parylene D molecule condenses and polymerizes more quickly than the other two, restricting its crevice penetration ability.

Parylene is applied dry via vapor deposition, which provides a very thin and uniform conformal coating. Coverage is excellent, and the coating reaches under the components, allowing the circuit to be operated even while immersed in water. Of particular importance for the preservation of electrical function in printed wiring assemblies (PWAs) are the epoxy gel-coat surfaces between conductor traces. Good adhesion here is essential to preserve electrical isolation between conductors in humid environments. In this situation, parylene's vapor deposition polymerization (VDP) condensation-diffusion-polymerization process offers a special advantage: because the parylene monomer is soluble in the epoxy gel-coat. Parylene polymer is formed under the surface of the epoxy, resulting in an interpenetrating polymer network, and an especially effective mechanism for adhesion [39].

Parylene has excellent chemical resistance, and is the only type of coating approved as resistant to chemical decontamination agents. Parylenes are much less permeable to moisture than other types of coatings. Parylene has dielectric properties that either meet or exceed those of other coatings at one-tenth the thickness or less; therefore, much less material is required for the same electrical performance [39]. The disadvantage is that it requires specialized and expensive equipment to apply. Repairing the coating requires special training. Because parylene C melts at 290°C recoating following field repairs must be done with another material, typically silicone.

13.1.6 Material Properties

Table 13.1 compares various conformal coatings in terms of their material properties. The available characteristics of conformal coatings are limited. There is often a potential for misleading assumptions when only one value is given for a particular parameter. Conformal coatings are polymers and their characteristics vary with temperatures and humidity. These variations can be quite significant if glass transition temperature is exceeded [104].

Due to the property limitations of a single coating material, a bilayer structure of two coatings may be used. This is especially used for the encapsulation of Micromechanical system (MEMS) devices in avionic applications. For example an interface-adhesion-enhanced bi-layer conformal coating has been developed for this purpose. It consists of first an application of silicone to planarize the MEMS surface due to its stress relief, shock/vibration absorber, and durable dielectric insulation properties. To compensate for the deficiency of silicone in preventing engine fuel/oil contamination, parylene C is deposited afterwards. To improve the adhesion between the two coating materials, silane coupling agents are used as adhesion promoters [110].

13.2 Application Techniques

The six methods for applying conformal coatings to circuit card assemblies are brush, dip, spray, wave, vacuum deposition, and needle based selective dispensing systems. A brief process description of each method, a discussion of some advantages and limitations, and a summary of coating options is provided below. Regardless of the application technique, the board must be properly prepared prior to coating, including cleaning and any necessary masking. Improper cleaning can trap flux residues and other ionic contaminants under the coating. These contaminants can decrease the system reliability leading to poor circuit performance and/or early wear out failures. Moreover, if the coating material is improperly applied and coverage is incomplete, the circuit will be susceptible to moisture and contaminant ingression [27].

Theoretically, it is thought that the thicker the coating, the better is its humidity-barrier properties, because the amount of moisture penetrating a coating is inversely related to thickness. However, with solvent-based coatings, the probability that solvent volatiles will be entrapped in the cured coating is greater with thicker coatings. In thinner coatings these volatiles are more easily released. Most circuit board manufacturers use coatings 0.5 to 3 mils thick. Thicker coatings can also cause components such as glass diodes or glass-sealed resistors to crack because of the stresses from shrinkage of the coating when the solvent evaporates, shrinkage from polymerization during curing, or large differences in the coefficients of expansion between glass and the plastic coating [108].

Table 13.1 Conformal coating material properties [39, 103, 111, 112]

Property	Urethane	Silicone	Acrylic	Epoxy	Parylene N	Parylene C	Parylene D
Nominal thickness, mils	1-3	2-8	1-3	1-3	0.5-1	0.5-1	0.5-1
Volume Resistivity (Ω cm) @ 50% RH & 23°C	10^{12}-10^{17}	10^{11}-10^{14}	1×10^{15}	2×10^{15}	1.4×10^{17}	8.8×10^{16}	2×10^{16}
Dielectric constant @1 MHz	3.3-4.0	4.2-5.1	2.2-3.2	2.6-2.7	2.65	2.95	2.80
Dissipation factor @ 1 MHz	0.03-0.05	0.05-0.07	0.07-0.14	0.001-0.002	0.0006	0.013	0.002
Thermal conductivity (W/m°C)	0.17-0.21	0.04-0.12	0.12-0.25	0.15-0.31	0.12	0.082	-
CTE (ppm/°C)	45-65	100-200	50-90	60-90	69	35	30-80
Maximum useful temperature (°C)	121	121	121	204	200	200	200
Resistance to weak acids	None	None	Slight	Very little	Good	Good	Good
Resistance to weak alkalides	None	None	Slight	Very little	Good	Good	Good
Resistance to organic solvents	Good	Good	Good	Good	Good	Good	Good

13.2.1 Brush

Of the five application methods, brushing is the most rudimentary. The coating can be applied either manually or by machine, with a worker responsible for quality coating. Typically, the brush method is only used for touch-up and/or repair operations, small parts, and low-volume applications.

There are several inherent problems with the brush method. The first is that the applicator□ brush or swab□ can apply contaminants along with the coating material if not clean. Particles from a swab can dislodge and become trapped in the coating. Secondly, the area to be coated must be thoroughly cleaned. This is difficult when repairing or touching up previously coated areas. Removal of the old coating, by chemical, melting, or abrasion, can damage the coating at the edge of the repair area and introduce contaminants into the edges of the old coating. Thirdly, achieving the correct viscosity is very difficult. During touch-up it is almost impossible to insure that thickness tolerances are maintained, due to the possibility of overlapping the old coating.

Brush applications can be labor intensive, require special ventilation areas for workers, and are not normally environmentally friendly [27]. The coatings used with this method are generally air-dryable solvent-based or moisture curable. Thickness problems increase the chances for pinholes and/or voids in the coating.

13.2.2 Dip

The dip method has been used since the early stages of conformal coating technology. Dip coating can be applied either manually or by machine. Prior to dipping, critical areas on the boards, such as test points and connectors, need to be masked. Using wire hooks or quick release fasteners, the boards are attached to a conveyor and automatically or hand dipped. They are immersed vertically and then removed from the reservoir. Immersion and withdrawal rates, as well as viscosity, are critical to the thickness and the elimination of air bubbles. In some cases the application process takes place in a partial vacuum, which helps eliminate air bubbles and draws out trapped air pockets around and under components. Once the board is removed from the vat, the excess coating material drips off. The boards are cured by ultraviolet light or heating.

As with brushing, it can be difficult to control the uniformity of the coating thickness. Due to surface tension, excess coating may be on the lower side. The coating will also be thicker at the intersection of perpendicular angles, such as where leads attach to the substrate and component body, or where components are flush with the substrate. It can be difficult to assure coverage of pointed ends, soldered lead ends, and sharply angled edges [78].

Solvent-based coatings require special ventilation areas. When choosing a dip process, the following additional factors should be considered:

      ∞   Contamination in the tank from the surface of the circuit card assembly

∞ Specific gravity of the solvent (to monitor solvent loss due to evaporation)
∞ Using an inert gas blanket for moisture sensitive materials
∞ Excessive solvent dripping

13.2.3 Spray

Spray is the most common method of applying conformal coatings. Applicators can be as simple as a hand-held spray gun or as complex as an elaborate automated process. Masking of a selective area during coating process is required. Material usage and waste can be high, but is generally less than dipping.

In 1996 the Swedish Institute of Production Engineering Research (IVF) issued a report saying that the application methods used by the industry at that time gave poor coverage of component leads, especially when the coating is applied using spraying as the application method [78]. Several processes have been developed to address this issue, including multi-angled spray patterns and/or multiple spray passes, allowing each layer to cure between passes. This results in a gradual build-up of coating material without viscosity run-off or surface-tension thickening. These processes require close control of viscosity, patterns, angles, temperature, humidity, and the amount of time the spray is applied to the surface.

There are three types of spray coating techniques: conventional, selective, and jet. The conventional method uses either a fixed or articulated head configuration. The boards are typically masked and fixed to a conveyor system, with a liner to catch any overspray. The fixture allows the boards to be turned so that both sides can be sprayed [27].

The selective coating process has become the method of choice for many medium and high-volume users. Selective coating can virtually eliminate the masking and demasking process, providing excellent material utilization. The selective coating systems can apply both solvent-based and solvent-free coatings, and use sophisticated robotic systems to provide high consistency and transfer efficiency. The system uses airless spray guns to apply the coating to selected areas of the boards.

The jet spray technique is similar to micro-jet soldering and ink-jet printing operations. It eliminates the need for masking and overspray disposal, and is capable of attaining very thin and uniform coating that is virtually free of pinhole flaws. The one instance when spraying does not provide good coverage is when it needs to be applied to the underside of low-profile components, such as TSOPs and PGAs [27].

13.2.4 Wave

Three wave-coating techniques are currently being used in the industry: conventional, megasonic, and meniscus. Conventional wave coating, a well-defined and mature process, is similar to wave soldering. First, the critical areas of the boards are masked, and then the assembly is placed in a fixture that is passed through the coating wave. The coating fluid is forced through a rectangular orifice to form a wave that wets the surface of inverted boards. The volume flow rate, fluid temperature, and belt speed are set to achieve optimum coating. After removal from the wave, the boards are sometimes brushed to remove and recover excess coating material. Coating thickness is still difficult to gauge [28].

Megasonic wave coating uses a piezoelectric parabolic transducer immersed in the reservoir to direct radio frequency (RF) energy at the coating surface. This RF energy establishes a stable wave that is easier to control than a conventional wave, and the quality of coverage is also better. Both the conventional and megasonic wave coating processes are typically limited to bottom-side coating; the maximum component height is also restricted [28].

Meniscus wave coating creates a fountain-type wave that results in convex surfaces of the coating material at the leading and trailing edges. Typically used for flat surfaces, this technique can achieve good thickness uniformity and attain a layer thickness of one tenth of a micron [28].

13.2.5 Vacuum Deposition

Vacuum deposition is a solvent-free process that deposits a uniform, highly conformal, pinhole-free film of high dielectric parylene onto the boards. Application thicknesses typically range from 15 to 40 micrometers, but thickness on the order of 0.03 micrometers can be achieved [34]. When vacuum deposition is used, the boards are cleaned, masked, and, at times, treated with an adhesion-promoting agent. The bonds are then placed in a vacuum chamber, where the coating is deposited as a gaseous monomer that polymerizes on all the surfaces. The vacuum deposition process releases no emissions and requires no curing or disposal. Since polymerizing gas does not exhibit liquid properties, such as bridging or meniscus forming, there are no mechanical stresses on the assembly either during or after deposition.

Vaccum deposition process is normally used when high reliability and light weight are the predominant concerns. Since parylene is transparent and applied in thin layers, visual inspection is difficult. However, with a fluorescent activation process, exposure to an ordinary black light makes inspection for complete coverage easier.

13.2.6 Needle Based Selective Dispensing Systems

Needle based selective dispensing systems offer greater efficiency and precision and consistent edge definition by putting the conformal coating material exactly where it is required. This technique can eliminate contamination risks, minimize concerns with viscosity variability, and avoid the costs of frequent purges and discards of unused fluid. However even with high-speed robotics, the dispensing methods are most appropriate for off-line, touch-up work in which accuracy and consistency are important but throughput and material thickness are uncritical [113].

13.2.7   Summary of Application Techniques

The various application methods are summarized in Table 13.2.

Table 13.2 Summary of Application Techniques. With permission from the Institute for Interconnecting and Packaging Electronic Circuits.

Features	Brush	Dip	Spray	Wave	Vacuum	Needle Dispensing
Coating promotion	Never	Never	Rarely	No	Usually	No
Transfer efficiency	Modest	Poor	Poor	Good	Excellent	Excellent
Thickness control	Poor	Poor	Good	Good	Excellent	Good
Bridging	Poor	Poor	Modest	Modest	None	Modest
Masking	Minimal	Difficult	Modest	Minimal	Very difficult	Not required
Environmental performance	Poor	Good	Modest	Good	Excellent	Modest

13.3 Curing Methods

Many solvent-based coatings have two stages of curing. The first stage is the evaporation of the solvents, which occurs at room temperature or with applied heat. The second stage involves the absorption of oxygen molecules and is a slow process taking up to 1000 hours at room temperature. The oxygen combines with the polymeric links and completes the chains allowing the coatings to reach their full filtering capabilities. When these types of coatings are used to underfill between the components and the substrate, the surface area exposed to the atmosphere is greatly reduced and the polymeric links may not be completed. Generally, the oxygen absorption rate cannot be sped up with heat.

The various methods for curing conformal coatings can be divided into four categories: ultraviolet (UV) light exposure, two-component mixing, thermal curing, and moisture curing.

13.3.1 Ultraviolet

Ultraviolet-cured conformal coatings were developed in the early 1980s after the Environmental Protection Agency (EPA) and the Occupational Safety and Health Administration (OSHA) passed strict regulations against volatile organic compound (VOC) emission levels. These coatings were solventless, reducing air pollution and saving energy by eliminating drying or curing after coating. They were successfully developed by creating acrylated oligomers, such as epoxy acrylates, urethane acrylates, polyester acrylates, or combinations of them. The greatest advantage of UV is the quick rate at which curing takes place. One of the problems is that it is a line-of-sight process, and shadowing will block curing. One possible solution is incorporating of multiple cure mechanisms into the process, such as UV and moisture.

Another problem with the UV curing process is that it is subject to cure inhibition at the coating-air interface. This occurs when the coating comes into contact with various sulfur, amine, or organometallic compounds sometimes found on boards as residual contaminants from de-molding or solder fluxing. In this case, oxygen in the air reacts competitively with the free radicals generated during the cure [27].

13.3.2 Two-Component

The two-component cure process allows the mating of two materials, often with an atomized spray. The main advantage in using the two-component cure method is that it offers a large degree of freedom in defining the cure speed and pot life. The main disadvantages of choosing the two-component method are that it requires the user to inventory two items instead of one and that the user is responsible for maintaining a proper mixture ratio while using this method. Waste can also be more in this process.

13.3.3 Thermal

Heat will generally increase the rate of cure if thickness variations occur; this method will amplify the uneven cure rate tendency. Control of the ramp up, dwell, and ramp down times, as well as the temperature levels, is necessary to prevent stress damage to components and solder joints plus cracking and delamination of the coating. Because the surface of solvent-based coatings will harden first, evaporating solvent vapors can become trapped, creating pinholes, bubbles, and voids. The best approach to avoid these problems is to lay down multiple layers of very thin coatings, using the thermal process to cure each layer during the buildup.

13.3.4 *Moisture*

The advantage of using moisture cure materials is that they offer rapid cure times. The disadvantages are that it is difficult to control the coverage area and that the equipment tends to easily clog, requiring frequent cleaning.

13.4 Coating Removal

There is always a probability that one or more defective components or solder joints in a printed circuit board will have to be repaired. Therefore it is important that the module be reworkable. Hence in addition to meeting the numerous engineering and manufacturing requirements, the coating must be easily removable so that defective components may be replaced. The removal technique must be one that does little or no damage to adjacent components, surfaces, and markings [108].

Various methods for removing conformal coatings are discussed in this section, along with some of the advantages and disadvantages of the processes. While each of the coating types has properties that make it a good choice for a particular application, attention must be paid to the techniques used for their removal.

The major ways to remove a conformal coating include chemical, mechanical, thermal, and abrasive. However laser and plasma etching techniques are also used for coating removal. The frequency of their use in the electronics industry is given in Figure 13.1.

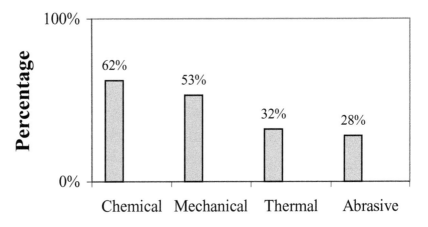

Figure 13.1 Removal methods used in the electronics industry [11]. (The percentages equal more than 100 percent because companies use combined techniques to remove their coatings.)

13.4.1 Chemical

The chemical technique is the most popular for removing conformal coatings. However, a perfect solvent for all applications does not exist. The first task in using this technique is to determine the composition of the coating being removed, either from the product literature or the manufacturer. When choosing a solvent, the following criteria should be examined:

∞ Does it remove the coating quickly and completely?

∞ Does it selectively remove the coating, while not damaging or adversely affecting the substrate and/or components or devices?

∞ Is it safe to work with?

∞ Is it environmentally acceptable?

There are four problems that are usually encountered in removal with solvents:

Solvents do not dissolve the plastic. The plastic can absorb large amounts of solvent and swell; and the softened material may then be removed by mechanical means.   The solvent cannot be localized; usually the entire assembly must be immersed, which presents the risk of damaging other areas. Swelling of the plastic can generate high stresses that may then induce other types of failure.

The time it takes to completely remove a conformal coating varies between fifteen minutes and twenty-four hours, depending on the amount and type of coating that was applied. The typical time for removal is between one and three hours. The use of ultrasonics or any agitation of the solvent will speed up the process. In general, the board should be checked periodically as the coating is removed. Occasionally, light brushing or wiping is required to remove bits of coating. To be certain that all traces of coating are removed, the board should be washed thoroughly with alcohol, rinsed with deionized water, and dried. To ensure no ionic materials are left on the board, the water should be checked after rinsing with a conductivity meter [32].

To remove a small area of coating to access a few components or traces, spot removal can be done with a high viscosity solvent or a gel and applied with a brush or cotton pad. When the coating has been removed, the area should be cleaned as described above. All the solvents discussed in this section are available in gel form.

Many solvents, with varying speeds and selectivity, can be used to remove urethane coatings. Methanol-base/alkaline-activator solvents are the most selective and the most popular. Ethylene glycol ether-base/alkaline-activator solvents are the fastest and least selective. For additional information, refer to Table 13.3 [26].

Table 13.3 Removal times of certain urethane coatings [26]

Urethane	Thickness	Solvent (1) Methanol-base/alkaline-activator solvents (2) Ethylene glycol ether-base/alkaline-activator solvents	Removal Time (Hrs)
Conap CE-1155	0.004	Either Type	1.5-2.5
Conap CD-1155-35	0.004	Either Type	1.5-2.5
Conap CD-1164	0.004	Either Type	1.75-2.5
Conap CE-1175	0.004	Either Type	1.0-2.0

Two types of solvents are effective in removing silicone coatings. These solvents, when not contaminated by water, will not attack epoxy-glass circuit boards, their components, metals, or other plastics. The fastest and most popular of these solvents is a methylene-based system. The alternative is a hydrocarbon-based solvent□ slower but more selective and recommended for spot removal. Typically, most silicone coatings will require fifteen minutes to an hour to be completely removed [26], but some of the newer silicones will not come off with any chemical. The operator should check with the manufacturer before using this technique. For additional information, refer to Table 13.4.

Table 13.4 Removal times of certain silicone coatings [26]. With permission from the Institute for Interconnecting and Packaging Electronic Circuits.

Silicone	Thickness	Solvent	Removal Time
Chemitronics Konform	0.003	Either Type	15-30 minutes
Conap CE-1181	0.006	Either Type	15-30 minutes
Dow Corning 1-2577	0.008	Either Type	15 minutes-1 hour
Dow Corning Hipec 3-6550	0.005	Either Type	15-45 minutes
Dow Corning Hipec 648	0.002	Either Type	15 minutes-24 hours
Dow Corning Hipec 01-4939	0.005	Either Type	15 minutes
GE SR900	0.004	Either Type	30 minutes

Chemically removing acrylics with highly flammable solvents is no longer acceptable. A relatively safe alternative using butyrolactone has been developed, and most acrylic coatings can be removed within an hour of soaking. For additional information, refer to Table 13.5.

Table 13.5 Removal times of certain acrylic coatings [26]. With permission from the Institute for Interconnecting and Packaging Electronic Circuits.

Acrylic	Thickness	Solvent	Removal Time (Hrs)
Conap CE-1170	0.007	Butyrolactone	1
Conap CE-1171		Butyrolactone	

Removing epoxy coatings from the entire board is nearly impossible because solvents cannot distinguish between the epoxy coating, the epoxy glass board, and epoxy-coated components. However, spots can be removed with a cotton swab if carefully done. The solvent to be used is a methylene chloride base and acid activator.

Parylene coatings can be chemically removed by immersion in a tetrahydrofuran base. The board should then be cleaned and dried. The coating can be peeled off the board with tweezers [26].

Because of the complicated structure of UV-cured coatings, chemical removal can be very complicated. While different manufacturers' coatings may use the same type of oligomer, their composition may vary enough to require different cleaning solvents. In some cases, a combination of solvents is required; in others, no solvent is suitable. Removal time varies from fifteen minutes to twenty-four hours, depending on the type of coating, coating thickness, and type of solvent used. For additional information on chemical removal of UV-cured coatings, refer to Table 13.6 [26].

13.4.2 Mechanical

Mechanical techniques for removing conformal coatings include cutting, picking, sanding, and scraping. Since the coatings are removed as solids, very little clean-up is involved using this technique [37]. However, most conformal coatings are very tough and abrasion-resistant, making the probability of damage to the board high.

Urethanes and epoxies are extremely hard after they are dry, so mechanical techniques are not recommended. Since silicones are more elastic than other types of coatings, the mechanical techniques can be used. Acrylics and parylenes can also be scraped off fairly easily. Most UV-cured coatings are a cross between polyurethane and acrylic; in this case, coating thickness will determine if this technique should be used. This technique must be used carefully so the board and/or components are not damaged [37].

13.4.3 Abrasive

Small sandblasting machines are used to remove conformal coatings. These machines introduce a cutting medium into a compressed air stream and eject through a handheld nozzle. The nozzle is directed where the coating is to be removed. The problem with these machines is that they generate static electricity, which can cause electrostatic discharge (ESD). Nevertheless, several types of cutting media are used in the industry, which take into account the ESD problem.

As long as the ESD problem can be controlled, the abrasive technique is good for most conformal coatings. However, many companies do not want to risk damaging their boards and components, so they find other methods to remove the coating.

Table 13.6 Removal times of certain UV-cured coatings [26]. With permission from the Institute for Interconnecting and Packaging Electronic Circuits.

Coating Type	Thickness	Solvent	Removal Time
Urethane Acrylates			
Dow Corning X3-6765	0.007	Methanol base/acid activator	4 hrs
Dymax Multi-Cure 984	0.006	Dimethylformamide base	15 min
Dymax Multi-Cure 984	0.006	Methylene chloride base/acid activator	15 min
Dymax Multi-Cure 984	0.006	Methanol base/acid activator	1 hr
Dymax Multi-Cure 984F	0.006	Dimethylformamide base	15 min
Dymax Multi-Cure 984F	0.006	Methylene chloride base/acid activator	15 min
Dymax Multi-Cure 984F	0.006	Methanol base/acid activator	1 hr
Dymax Multi-Cure 984RF	0.006	Dimethylformamide base	15 min
Dymax Multi-Cure 984RF	0.006	Methylene chloride base/acid activator	15 min
Dymax Multi-Cure 984RF	0.006	Methanol base/acid activator	1 hr
Loctite Shadowcure 361	N/A	No suitable solvent	N/A
W/R Grace Amicon UV-920	N/A	Methylene chloride base/acid activator	2.5 hrs
WR Grace Amicon UV-920	N/A	Ethylene glycol ether base/alkaline activator	24 hrs
Acrylated epoxy urethane			
DuPont Quickcure B-565	0.003	Dimethylformamide base	15 min
DuPont Quickcure B-565	0.003	N-methylpyrrolidone base	1.5 hrs
DuPont Quickcure B-565	0.003	Methylene chloride base/acid activator	15 min
DuPont Quickcure B-565	0.003	Butyrolactone base	45 min
DuPont Quickcure B-566	0.003	Dimethylformamide base	15 min
DuPont Quickcure B-566	0.003	N-methylpyrrolidone base	1.5 hrs
DuPont Quickcure B-566	0.003	Methylene chloride base/acid activator	15 min
DuPont Quickcure B-566	0.003	Butyrolactone base	45 min
Silicone			
Dow Corning X-4013	0.011	Methylene chloride base/acid activator	15 min
Dow Corning X-4013	0.011	Butyrolactone base	45 min
Dow Corning X-4013	0.011	Methylene chloride base/acid activator	15 min
Dow Corning X-4013	0.011	Ethylene glycol ether base/acid activator	45 min
Dow Corning X3-6760	0.004	Methylene chloride base/acid activator	30 min
Dow Corning X3-6760	0.004	Ethylene glycol ether base/acid activator	1.25 hrs
Dow Corning X3-6760	0.004	Hydrocarbon base/acid activator	1 hr

∞ *Aluminum oxide* is an aggressive medium that can cut right through a printed circuit board. Typical ESD readings range from 500V to 1000V.

∞ *Biological media*, such as wheat starch and walnut shells, are not as aggressive but they usually leave a residue that must be cleaned off prior to recoating. However, these media are not ESD friendly, generating more than 25,000V in many applications.

∞ *Sodium bicarbonate* is a popular medium, but it generates high ESD levels and must also be thoroughly cleaned off the board before reapplying a coating.

∞ *Plastic cutting media* have the lowest ESD generating levels and, if formulated correctly, are also recyclable.

For a 0.004-inch urethane coating on a single component, removal time is less than a minute. With a greater thickness, the time increases to only two or three minutes. Removing the coating on an entire board depends on the size of the board and the thickness of the coating.

For silicones, removing coatings up to 0.020 inches thick is extremely easy. Thicker silicones can be more difficult because the media will bounce off the coating [39].

For spot removing parylenes, it is recommended that the area be masked so that a clean edge is formed when the coating is reapplied.

13.4.4 Thermal

Thermal techniques are the least recommended of all removal techniques. They include using a heat gun or soldering iron to burn through the coating. Most coatings require a very high temperature and/or long exposure times that can cause discoloration, leave residues, and adversely affect solders and other materials. Thermal removal can lift surface-mount pads from boards and also damage temperature-sensitive components. Extreme caution is necessary when burning through conformal coatings, because some emit highly toxic and hazardous vapors [37]. For example, urethanes or UV-cured conformal coatings emit a toxic gas that can harm workers. The only way to use this method on urethanes is to decompose the coating to carbon dioxide and water. Similarly, silicone coatings can only be spot removed because of the fumes; the manufacturer should be consulted before using this technique.

Thermal methods are the most popular way to remove acrylic and parylene coatings. Epoxy coatings can be cautiously removed by pyrolizing the coating material.

13.4.5 Laser

When electromagnetic radiation from a laser is absorbed by organic polymers, the polymer molecules become excited. When the radiation beam has enough power to heat the polymer sufficiently to produce a vapor pressure then the polymer is volatilized. If the photon energy is sufficient to break chemical bonds, volatile compounds are produced. Thus laser radiation can remove conformal coatings from circuit boards by two mechanisms:

∞   Coating photodegeneration of the coating
∞   Chemical photodegeneration of the coating

Adequate radiation exposure time is required to complete the coating removal from the desired area by either mechanism. Less laser beam power is required for photodegeneration than for ablation. Ultraviolet lasers are required for photodegeneration whereas other lasers are used for melting and vaporization. By focusing the laser radiation beam, small coating areas can be removed. By sweeping the beam, the entire coating can be removed by either process. With either technique, care must be taken to ensure that there is no damage to the circuit board and its electronic components [107].

13.4.6 Plasma Etching

Plasma coating removal requires a vacuum system, an electromagnetic radiation source, and a gas supply. Either radio frequency (103-109 Hz) or microwave frequency (109-1012 Hz) sources can be used. Plasma etching mechanisms are divided into four categories:

∞ Sputtering mechanism
∞ Chemical mechanism
∞ Ion enhanced energetic mechanism
∞ Ion enhanced inhibitor mechanism

In sputtering mechanism ions mechanically eject coating materials at low pressure whereas thermalized neutral gaseous radicals react with coating material to form volatile products in the chemical mechanism. The ion enhanced energetic mechanism is characterized by little or no intrinsic surface reaction with neutral radicals until energetic ions increase the reactivity of the coating producing chemical reactions, which form volatile molecules from the coating. By depositing an inhibitor to exclude etchant in working areas, small localized areas can be etched [107].

Table 13.7 Recommended removal techniques based on coating type. With permission the Institute for Interconnecting and Packaging Electronic Circuits.

Coating Type	Spot Removal	Whole Board	Clean Up
Urethanes	Abrasive Chemical Mechanical	Chemical Abrasive	Mechanical Abrasive Chemical
Silicone	Abrasive Chemical Mechanical Thermal	Abrasive Chemical	Mechanical Abrasive Thermal Chemical
Acrylic	Abrasive Mechanical Thermal Chemical	Abrasive Chemical	Mechanical Thermal Abrasive Chemical
Epoxy	Abrasive Chemical	Abrasive	Abrasive Chemical
Parylene	Abrasive Thermal Mechanical	Abrasive Chemical	Mechanical Thermal Abrasive
UV-cured	Abrasive Mechanical Chemical	Abrasive Chemical	Mechanical Abrasive Chemical

13.5 Reliability of Conformal Coatings

Early in the 1990s a landmark study was commissioned by various European assembly manufacturers to study the reliability of conformal coatings. The Swedish Institute of Production Engineering Research (IVF) issued a series of reports based on the results of this study [78-83]. The study demonstrated some serious problems with the methods used to test coatings. In particular it was found that existing specifications, tests, and standards were far too benign and were performed primarily on blank coupons. Under these circumstances it was impossible to determine the reliability of the coatings.

Existing specifications, tests, and standards are primarily used to qualify the coating materials and have no relationship to the manufacturing process or the end use environment. It is currently up to the assembly manufacturer to develop and perform the tests necessary to determine coating reliability. The tests require verification of the cleanliness of the board and components, proper coverage, plus the ability to prolong circuitry operation in the end use environment. Testing specially designed dummy components and prototype assemblies using expected environmental conditions is the best method to verify the compatibility of the materials, processes and application requirements.

Pinholes are inevitable and often arise as the coating shrinks during curing. A high pinhole density increases the possibility of moisture or other contaminants migrating under the coating over time and can serve as a focus for crack initiation.

One test for determining coating integrity is inspection under a black light. This is not very effective since it shows only large holes or voids, even under magnification. This test is incapable of determining a coating thickness. A possible solution to determining coating quality is using test coupons with dummy components, prototypes, or samples of production units in an environmental chamber. By creating condensation on the coating surface and measuring the leakage currents when a bias is applied, one can determine the integrity of materials used in a coating. When moisture combined with pollutants reaches the board surface oxidation and corrosion of metal circuitry and component leads will begin. The bonding of the coating will tend to break down at the interface of the metals to the coatings. These are problems which can occur when there is no power applied to the circuitry. When power is applied, under these circumstances, electrochemical migration can begin leading to dendritic growth. To effectively determine the ability of the coating to filter pollutants, SIR testing must be done while the coated circuit is exposed to a high humidity polluted atmosphere with and without power supplied [107].

Another possible test of coating damage is rolling a moistened swab over the circuit coating and observing the operation of the circuit. Care must be taken in this test because leaving the moisture on the board too long could cause more contamination [85].

Accelerated life cycle testing is capable of demonstrating long-term reliability of conformally coated boards. However there are several problems with establishing parameters for accelerated testing. Coatings are organic polymers and thermal plastic by and large. The changes in physical characteristics, which occur with changes in temperature, are very poorly documented. This is especially true if temperatures exceed the glass transition temperature of the coating. Moreover, the complex synergistic chemical reactions that occur in the atmosphere and on, in, and under the surface of the coating are difficult to duplicate in a test chamber [107].

The single most impedance to conformal coating adhesion is surface contamination. The presence of ionic residues, oily materials, and particulates on board surfaces and components can result in corrosion, insulation breakdown, and poor conformal coating adhesion. Ionic contamination will cause vesication of the conformal coating. Oily materials and particulates will not allow most conformal coatings to adhere to the surface or substrate leading to peeling [107].

A test of the protection afforded by the parylene coating is to coat circuit board test patterns (as described in MIL-I-46058C) and subject them to insulation resistance measurements during a temperature-humidity cycle (MIL-STD-202, methods 106 and 302). In brief, this test consists of ten cycles (one cycle per day), with each cycle consisting of seven steps. The seven steps range from low temperature, low humidity (25°C, 50% RH) to more severe conditions (65°C, 90% RH). Resistance readings are taken initially and at the 65°C, 90% RH step for each cycle of the ten day test [103].

Many conformal coating formulations cure to hard, solid resin, offering excellent toughness against physical abuse and abrasion. However, the general trend toward miniaturization in electronics requires greater compliance from conformal coating materials to prevent undue stress on fine-pitch leads and very small components. Especially under repeat thermomechanical cycling, the coefficient of thermal expansion (CTE) mismatch between a stiff coating, the board substrate, and components can crack surface coatings and damage connections.

13.6 Environmental Concerns

In 1970, the US government passed the Clean Air Act, which gave the Environmental Protection Agency (EPA) the authority to set national air quality standards to protect against common pollutants, including materials that release volatile organic compounds (VOC) and ozone-depleting chemicals (ODC). All conformal coatings except parylene were solvent-based coatings until a decade ago but now due to the EPA and the Occupational Health and Safety Administration (OSHA) standards, there is an increasing trend towards the use of solvent-free and low-ODC/VOC conformal coatings. While these coatings are more expensive on a per-unit basis than solvent-based materials, much less volume is used. Because solvent-free coatings are 100 percent solids, they do not evaporate as part of the curing process. Solvent-based materials are typically 60-70 percent solvents, all of which is wasted during evaporation [106].

Environmental regulations, such as the Montreal Protocol and Clean Air Act, have had a significant impact on both coating materials and application methods, particularly with regard to control of volatile organic compounds (VOCs) and ozone depleting chlorofluorocarbon (CFC) compounds. VOCs are of primary concern because they react in the atmosphere to form ground level ozone, called smog. CFCs deplete earth's protective ozone layer in the upper stratosphere. Both VOCs and CFCs have been extensively used as solvent carriers. Manufacturers and suppliers of conformal coating materials have responded by developing non-solvent-based coatings and more environmentally acceptable methods of application, curing, and removal.

13.7 Selection Trade-Offs

A conformed coating must be selected on the basis of electrical, thermal, mechanical, and other pertinent properties, as dictated by requirements for circuit performance and characteristics, type and degree of environmental exposure, recoverability and consequences of failure. Five basic prerequisites for good coating performance include [85]:

∞   Surfaces to be coated must be free of contaminants and volatiles,

∞   Electrical contact areas must be adequately masked to be free of coating,

∞   Coating film should be capable of developing an adequate bond to the applied surfaces,

∞   Coating film should be completely cured and cover all intended areas at a specific thickness,

∞   The assembly coating, substrate, and (if applicable) the bare-board coating, should yield a dielectric system that complies with design requirements.

The practical aspects of production and repairability must also be taken into account. IPC-HDBK-830 gives the guidelines for design, selection, and application of conformal coatings [107].

All conformal coatings have inherent limitations. They generally prevent the formation of a continuous water film between electrical conductors, but can still yield a film that behaves like a semi-permeable membrane, permitting a degree of moisture penetration. The bulk volume resistance of the film will drop proportionally to the duration and severity of exposure to humidity, a typical and normal degradation that is reflected in surface insulation resistance (SIR) readings made during testing. Even so, the coating still protects the circuit from the effects of bridging water films that can cause signal changes like cross-talk or shorted circuits, metallic growth, or corrosion.

Designers should also be concerned with the following conformal coating factors:

∞   The dielectric constant of the coating material for high-frequency circuits must be known, with breakdown voltage and arc resistance for high-voltage signals, flame resistance, and fungus and reversion resistance for performance in humid climates.

∞ Mechanical and thermal properties, especially the CTE, are important considerations in managing mechanical stresses on solder joints and leads.

∞ Chemical properties, especially resistance to cleaning solvents, fuel, and hydraulic fluids, are sometimes required, and compromises must be made for producibility, repair, and rework.

If a solder mask must be used, it is best to specify solder mask over bare copper (SMOBC). Since different permanent masks vary in degree of adhesion to their conformal coatings, each combination must be evaluated for repeatability of performance. Plasma cleaning is an effective way to deal with solder mask/conformal coating adhesion.

Table 13.8 summarizes the benefits and the drawbacks of various conformal coatings. Table 13.9 compares some key properties of the conventional coatings with respect to the parameters that affect the proper function of electronic and electrical devices.

Table 13.8 Summary of advantages and disadvantages. With permission from the Institute for Interconnecting and Packaging Electronic Circuits.

Coating	Advantages	Disadvantages
Urethane	Good mechanical properties Good adhesion to components and board High chemical resistance Low moisture absorption Low viscosity Low cost Most widely used	Stable only up to 135℃ Flammable Difficult to rework Unsuitable for high-frequency circuits
Silicone	Excellent electrical properties Mechanically tough and flexible Resistant to high temperatures Low moisture absorption Low stress after cure High CTE	Low resistance to hydrocarbons Poor adhesion High cost
Acrylic	Excellent moisture resistance Good dielectric properties Easy to apply Fast drying Easy to rework Suitable for automated, high-volume production Ultraviolet curable modification available	Poor chemical resistance Poor mechanical abrasion resistance
Epoxy	Excellent chemical resistance Excellent mechanical properties Excellent wetting for application Good adhesion Acceptable moisture barrier	Compliant buffer needed Coating for low stress rework is difficult Short shelf life
Parylene	Penetrates hard-to-reach places Excellent chemical properties Excellent mechanical properties Excellent electrical properties Good adhesion Excellent moisture resistance Excellent control in manufacture	Masking of non-coated areas difficult Difficult to rework Needs expensive equipment to apply Low thermal stability

Table 13.9 Rating comparison of conformal coatings

Property	Urethane	Silicone	Acrylic	Epoxy	Parylene
Uniform, very thin, conformal layer	G	G	G	G	E
Low stress, pin-hole free layer	M	M	M	M	E
Dielectric properties	M	VG	G	M	E
Physical strength	VG	M	G	VG	E
Flexibility	VG	VG	M	L	VG
Wear and abrasion resistance	VG	L	M	VG	E
Coefficient of thermal expansion	M	L	G	VG	E
Water absorption	G	L	G	VG	E
Chemical, solvents, fungus resistance	VG	M	L	VG	E
Barrier to moisture, gases, liquids	G	M	VG	VG	E
Adhesion to substrates	G	M	VG	VG	G
Repairability	G	M	VG	L	G
No contaminating ingredients	G	L	G	G	E
Particle immobilization	L	L	L	L	E
Performance under humidity	G	G	G	G	G
Performance under water	L	M	L	L	E
Overall application characteristics	G	M	E	M	M
VOC exempt	M	M	M	M	E
Convey or processing capability	L	L	E	L	N/A
Affordability	E	L	G	G	L

E= Excellent, VG= Very Good, G= Good, M= Moderate, L= Low, N/A= Not Applicable

GLOSSARY

Abietic acid: a terpene carboxylic acid forming, along with some isomers, about 95 to 97 percent of the mass of wood rosin. Formula-$C_{20}H_{30}O_2$.

Abrasive: Abrasives are widely used in printed circuit manufacture. The most usual types are nonwoven synthetic mat flap brushes, nonwoven synthetic mat disc brushes, pumice slurry, and abrasive synthetic fiber brushes. All abrasive cleaning is contaminating, as particles are implanted in the metal. It is also one of the principal causes of dewetting. Abrasives may also be used for removing certain conformal coatings.

Absorption: a physical effect whereby a substance in a fluid phase will penetrate a porous surface in a solid phase by capillary forces.

Acetone: The simplest ketone is CH_3COCH_3, also known as 2-propanone and dimethylketone; a strong inflammable hydrophylic solvent.

Acetonitrile: More properly called methyl cyanide, this is a highly toxic polar solvent with a wide spectrum of solubilization, used for extraction testing.

Acid: a complicated concept of molecular chemistry. By simplification, can be defined as a substance that either directly or ionogenically can dissociate in water to produce one or more hydrogen ions.

Acid acceptance: amount of acid solute that a chlorinated or fluorinated solvent will accept before decomposing into strong mineral acids.

Acid index or number: the relative acidity of a compound expressed as the number of milligrams of potassium hydroxide required to neutralize one gram of the dry weight of the compound. In aqueous media, the neutralization is determined electrometrically, but under anhydrous conditions a chromatic indicator is used.

Acrylic resins: synthetic resins used for conformal coatings, based on the same family as polymethylmethacrylate resins; characterized as medium to good for most purposes, except as regards chemical resistance.

Active carbon: charcoal, derived from various organic sources, activated at high temperature to make it highly adsorbent; used for filtering fluids to rid them of certain undesirable organic products.

Adsorption: a physico-chemical effect by which an atomic or molecular bond is created at the interface between two substances.

Alcohol: any hydrocarbon with one or more hydroxyl groups (OH) bound to a carbon atom. The simplest is methanol (CH_3OH), followed by ethanol (CH_3CH_2OH), e.g., and so on. Glycols are alcohols with two hydroxyl groups (i.e., ethylene glycol is $OHCH_2CH_2OH$), and polyols have three (glycerol).

Aliphatic: an organic substance that has a linear molecular structure, as opposed to a ring structure.

Alkali: *see* Base.

Alloy: a mixture of two or more elements, of which the major components are metallic. An alloy is often a solid solution of intermetallics.

Amine: basic organic compounds whose molecular structure is similar to ammonia (NH_3), in which one or more of the hydrogen atoms is replaced by hydrocarbon groups. Used as saponifiers for solubilizing rosin and other carboxylic acids.

Amino acid: an organic acid containing both a carboxyl (-COOH) and an amino (-NH$_3$) group in its composition, found naturally in animal and vegetable proteins. Frequently used as an active principal in fluxes, either by itself or as its hydrochloride.

Anhydride: a substance formed by eliminating one or more molecules of water from each molecule of another substance by dehydrolization.

Anion: a negatively charged ion.

Antifoaming agent: a product added to a solution to prevent it from foaming unduly by changing the surface tension; usually derived from octyl alcohol.

Antistatic plastics: Many plastic films and foams used for packaging electronic components and assemblies are rendered conductive to prevent the accumulation of electric charges that could destroy sensitive semiconductors. This is achieved by several methods, e.g., the use of conductive carbon fillers, hygroscopic plasticizers, etc. Some of these methods produce dangerous contaminants that may transfer to the product wrapped in them.

Aqueous cleaning: a method of decontamination in which water, with or without suitable additives, is used as the solvent. This is the ideal method for removing dangerous ionic or ionogenic contaminants, but is less satisfactory for removing hydrophobic ones unless the additives used are very carefully chosen.

Aromatic: an organic substance that contains at least one benzene ring in its molecular structure.

Atom: the smallest part of an element that is indivisible by ordinary chemical change.

Atomic weight: the relative weight of an atom based on a carbon 12 atom (= 12).

Autoelectrolytic corrosion: electrolytic corrosion in which the electric current is generated voltaically between the metal and another metal, between the metal and an alloying inclusion, or with a single metal and different ion concentrations.

Avionics: branch of electronics devoted to the instrumentation for aircraft and space applications.

Bactericide: a product used to kill living bacteria.

Bacteriostat: a product used to prevent bacteria from finding a medium on which to thrive.

Base: a complicated concept of molecular chemistry. By simplification, can be defined as a substance that either directly or ionogenically can dissociate in water to produce one or more hydroxyl ions.

Bipolar solvent: synonymous with polar solvent. By definition, all polar solvents have two poles.

Bleed-through: literally, permeation of resin through pinholes in the copper cladding of laminate during pressing. Can also be applied to contamination carried through epitaxial micropores in the copper by solvent evaporation during pressing.

Blistering: local lifting of internal layers of a laminate or of a surface coating, due to lack of good adhesion.

Blow holes: craters formed in solder joints due to out-gassing.

Brighteners: commercial products specifically designed for cleaning metal surfaces, removing oxides or other tarnish products, and rendering the surface reactive or solderable.

Butanol: any of four isomeric alcohols with the formula C_4H_{10}.

Buttercoating: an additional layer of resin added to the surface of a laminate to increase the thickness between the outside and the glass fibers, for any reason. Ordinary laminate is not buttercoated. Some laminates for the additive process are buttercoated with an epoxy butyl copolymer that can be etched.

Capillary effect: the penetration of liquids into small spaces by surface tension.

Carboxylic acid: an organic acid containing one (monobasic), two (dibasic), or more carboxyl groups (-COOH). Monobasic and dibasic aliphatic types (fatty acids) are frequently used as active principals in fluxes. Soaps are the organometallic salts of monobasic acids.

Catalyst: a substance that can accelerate or decelerate a chemical reaction while remaining unchanged.

Cation: a positively charged ion.

Cavitation: the formation of voids in a solvent due to a sudden depression. Cavitation is an essential part of ultrasonic cleaning.

Ceramics: inorganic nonmetallic products formed by various means and fired; including glasses, enamels, porcelains, earthenware, steatites, alumina, etc.

Chelation: in practical terms, a process by which an insoluble product is "imprisoned" within the structure of a soluble one, the resultant complex remaining soluble. It is used for solubilizing insoluble contaminants in cleaning processes, notably organometallic salts.

Chemical formula: the combination of symbols and the respective ponderal proportions to represent the composition of a molecule or a compound. *See* Empirical, Molecular, and Structural formula.

Chip carrier: any one of a number of designs of housings for connecting semiconductor chips to a printed circuit.

Chlorinated solvents: a halogenated solvent with at least one chlorine atom.

Chromatography: A number of analytical techniques are grouped under this heading; useful in chemical laboratories. The method that is successful for contamination testing is ion chromatography.

Citric acid: a tribasic carboylic acid occasionally used in water-soluble fluxes. It decomposes at soldering temperature into dibasic aconitic acid and water.

Colloid: a dispersion of microparticles in a gaseous, liquid, or solid phase; particles are so small that their surface properties alter the physical properties of the phase and remain constant in time.

Colophony, colophonium: synonyms for rosin.

Combination cleaning: flux removal by successive cleaning in organic solvents and aqueous cleaners.

Complex: a compound with imperfect valence bonding but in which the entity formed is stable and has properties that may or may not be independent of the component molecules.

Compound: a chemical substance that is of uniform composition and structure and consists of molecules containing at least two different kinds of atom.

Conchoidal fracture: the manner in which a hard, noncrystalline substance breaks, usually with a curved surface, often with concentric arcs in relief. Its form is frequently reminiscent of that of a seashell, hence the name.

Condensation polymerization: a form of polymerization in which a liquid by-product is generated and has to evaporate before the process is completed.

Condensation soldering: *see* Vapor phase soldering.

Conductance: specific conductance and conductivity.

Conformal coating: a thin protective coating of more or less constant thickness whose form follows closely that of the assembly underneath it.

Contaminant: any single product causing contamination.

Contamination: in the context of this document, any undesirable product on the surface of an electronic component or assembly, adsorbed on the surface or absorbed into the surface, which may cause problems either during the manufacturing process or by reducing the reliability of the finished assembly during its expected lifetime.

Contamination control: an approach by which a whole series of processes is checked at each stage for the introduction of new contaminants, which are eliminated before the next stage.

Copolymer: a synthetic polymer composed of two polymers molecularly mixed into single chains. It is not a simple physical mixture of two polymers.

Copper-clad laminates: base material for the manufacture of printed circuit boards by a subtractive process. Consists of an insulating substrate (glass fiber-reinforced epoxy resin) to which are applied, on one or both sides, sheets of treated electrolytically deposited copper foil.

Cross-linking: joining together chains of linear polymers to form a thermosetting resin, by attaching molecular groups across and between the chains in three dimensions. The process of cross-linking is called curing and may be a purely thermal reaction or initiated by light or other energy sources. Curing is often highly exothermic. Most laminates, solder resists, and permanent conformal coatings are made from cross-linked resins.

Crystal: a solid whose form is determined by the arrangement of atoms, ions, or molecules within it.

Curing: process of cross-linking.

Degassing: a fault process by which a gas or vapor is generated or released from any source to condense or react elsewhere, often forming a contaminant. Degassing is particularly dangerous in avionics, where flux residues or other contaminants may evaporate under reduced atmospheric pressure to redeposit themselves where they are even less welcome. The choice of fluxes for spacecraft is therefore particularly critical, many commercial ones being prone to degassing and redepositing themselves over the whole interior of a closed assembly.

Deionized water: water whose impurity ions have been exchanged for hydrogen and hydroxyl ions with a low conductivity. Deionized water may contain nonionic impurities.

Deionizer: *see* Ion exchange resins.

Deliquescence: the property of absorbing sufficient atmospheric humidity that its equilibrium point is in a liquid phase (solution). *See* Hygroscopic.

Denaturant: product added to ethanol to give it an unacceptable taste; must not be separable from it by any means.

Dendrite: a tree-like growth. In the context of this document, it refers to the growth of electrolytic whiskers between conductors of, for example, printed circuits.

Dephlegmator: a condenser used for reflux generation.

Depolymerization: the opposite of polymerization, also known as reversion or resin reversion.

Desiccator: a system for drying. More specifically, a means of removing water from organic hydrophobic solvents by means of a molecular sieve. Also, a piece of glass laboratory equipment in which a desiccant or other product is placed, with a cover over it, the whole being isolated from the atmosphere.

Detergent: synthetic product, with an action similar to soap, with long chain molecules, opposite ends of which are lipophilic and hydrophilic. The most usual types are termed anionic detergents, as their solutions are ionic with large anions present.

Dewetting: any phenomenon by which the liquid on a wet surface will be drawn back into discrete globules, analogous to water on greasy glass. In particular, the term applies to a similar effect when solder on a metallic surface apparently wets initially and then retreats. The most usual cause is myriad micropoints of contamination over the metallic surface, such as may be caused by bleed-through or abrasive particles implanted in the metal. In this case, the only cure is decontamination of the surface. Dewetting may also be caused by contaminants in the solder alloy. As a general rule, an increase of flux activity will not reduce dewetting, although a reduction of soldering time may help.

Dielectric: a substance that theoretically contains no free electrons and is thus the perfect insulator. In practice, the term is almost synonymous with insulator, particularly those used for separating the electrodes of a capacitor.

Dielectric constant: the ratio, related to a perfect vacuum, by which a dielectric increases the electrical capacity of the space between two electrodes. It is related to the number of free ions or electrons in the dielectric, but is unrelated to dielectric strength or surface or volume resistivities. Water has a very high dielectric constant or permittivity (approximately 80) as do other polar solvents (methanol:30, ethanol-25, propanols-20, butanols-15), decreasing in the order of their polarity. Epoxy resins can have excellent electrical properties but a fairly high dielectric constant, due to the presence of isolated salt molecules in the resin mass.

Dielectric strength: the voltage required to break down a given thickness of dielectric.

Diisocyanate: *see* Toluene diisocyanate.

Dipole: any electrical system where two poles have opposite polarities. A molecular dipole is asymmetric (*see* Polar solvent), and may be the vector sum of different ionic charges.

Dissociation: ionic or electrolytic dissociation is the process by which a neutral molecule is split into ions under the influence of an electrical field. The electrical field may be applied from an external source or may be that between the dipole of a polar solvent, e.g., water.

Dissolution: the act of a solvent dissolving a solute.

Distillation: the recovery of a single solvent or an azeotrope by boiling the dirty solvent and condensing the cleaner solvent. Reflux distillation offers a much better quality of condensate, at the cost of a lower thermal efficiency.

Drag-out: transferring contamination from a dirty solvent to a clean one as the liquid adheres to a part being cleaned, particularly under components and in crevices. Bare, drilled printed circuits are notoriously prone to drag-out.

Dry film resist: a photosensitive polymer that is heat-laminated onto a printed circuit during the manufacturing process. It is then exposed through a phototool and developed to remove all the unexposed product, the exposed material remaining in place.

Dyes: coloring materials added to put a color on a surface; the color is produced at the molecular level.

EDTA: ethylenediaminetetra-acetic acid. A common product whose salts are frequently used as chelating agents.

Elastomer: a polymer with rubbery qualities. Natural rubber is one.

Electrochemical series: a qualitative list of elements drawn up in the order of their electrochemical voltages or oxidation potentials. The greater the difference in oxidation potentials, the greater the risk of corrosion when two such elements are in electrical contact.

Electrolysis: the effect produced by passing an electric current through an electrolyte. At the cathode, the cations gain an electron to form atoms (usually metallic or hydrogen); at the anode, the anions lose an electron to form oxygen atoms which may combine with the electrode material to redissociate.

Electrolyte: an aqueous solution in which cations or anions other than hydrogen and hydroxyl ions are present and the conductivity of which is greater than that of pure water under the same conditions.

Electrolytic cleaning: anodic or cathodic cleaning of a metal in a special electrolyte. This is sometimes used prior to electroplating.

Electrolytic corrosion: corrosion of a metal in the presence of moisture and free ions due to the passage of an electric current.

Electrophoresis: the migration of charged particles in an electric field. Strictly speaking, electrolysis is a form of electrophoresis but, popularly, the term is generally applied to particles of greater than atomic size.

Electroplating: the cathodic deposit of a metal from an electrolyte containing ions of that metal.

Element: a substance that can neither be formed nor broken down by ordinary chemical change.

Elute: solution used for carrying the sample to the columns in various forms of chromatography.

Empirical formula: the chemical formula of a compound, determined experimentally and reduced to its simplest monomeric form.

Emulsion: a dispersion of fine particles of one liquid in another liquid that is immiscible with the first. The most common emulsions are either fatty bodies in water or vice versa. Generally, emulsions require either a violent mechanical action or an emulsifying agent, i.e., a detergent, to form or to remain stable.

Endothermy: the property by which a chemical reaction occurs, creating a depression of temperature.

Epitaxy: crystal growth as a continuum across a boundary. For example, copper plating to a crystalline copper surface may result in an epitaxial extension of the crystal and pore structure.

Epoxy group: An organic compound with an oxygen atom valence bonded to two adjacent carbon atoms is said to contain a epoxy group.

Epoxy resin: a synthetic polymer in which cross-linking of linear chains is achieved by modifying and polymerizing epoxy groups.

Ester: an organic compound produced by the reaction of an acid on an alcohol, with the liberation of water. The reaction is reversible and can only be completed if the water is eliminated. It usually requires heating to esterize most alcohols. Esters are often volatile sweet-smelling "fruity" liquids. A well-known example is amyl acetate, which has a strong odor of pears and is a solvent frequently used in nail-varnish removers.

Etching: the removal of copper during the subtractive process of printed circuit manufacture. Etching may, under certain conditions, leave insoluble contaminants that may become ionogens on the circuits. If they are not removed, these contaminants may become particularly dangerous.

Etch resist: product designed for locally preventing etching in order to generate the conductor pattern. According to the process used, etch resists may be organic (photoresists, dry film resists, or silk-screened resists) or metallic (electroplated tin, tin/lead, nickel, gold, etc.). In the case of organic resists, a subsequent stripping is an essential part of the process. If it is not carried out properly, the residues become contaminants. Over-stripping may attack the substrate, causing contamination absorption. Metallic resists are themselves frequently contaminated during the etching process and must be subsequently cleaned chemically.

Ether: an organic compound produced by the oxidation of an alcohol. Ethers are usually volatile liquids with a pleasant odor. The most common is diethyl ether, $CH_3CH_2OCH_2CH_3$, well known for its anaesthetic, solvent, refrigerating, and explosive properties.

Ethylamine: *see* Amine.

Eutectic: adjective describing the temperature or composition of a mixture of two (binary eutectic), three (tertiary eutectic), or more substances the melting point of which is at a minimum; it passes directly from a solid to a liquid phase at the eutectic temperature on heating and vice versa on cooling.

Exothermy: the property by which a chemical reaction occurs, creating an elevation of temperature.

Fatty acids: monobasic carboxylic acids.

Filaments: metallic growths from conductors. *See* Dendrites.

Flame point: the lowest temperature at which an inflammable product will ignite spontaneously, without application of an external flame.

Flash point: the lowest temperature at which vapor above an inflammable liquid will ignite under given conditions. Various standard measuring methods will give slightly different results.

Fluorinated solvents: a halogenated solvent in which at least one fluorine atom exists per molecule. Most fluorinated solvents contain chlorine as well as fluorine.

Fluorocarbon: an organic hydrocarbon in which at least one hydrogen atom has been replaced by a fluorine atom. It is incorrect to call fluorinated solvents also containing chlorine atoms fluorocarbons; the correct term is fluorochlorocarbons.

Flux: a product, usually liquid, used to provide the chemical reactivity necessary to reduce oxides, to provide an inert atmosphere to prevent reoxidation, and to reduce the metallic surface tension during a soldering operation.

Freeboard: in a vapor-phase solvent cleaner, the vertical distance between the top level of the solvent vapor and the opening, permitting the entry and exit of parts being cleaned. Minimum solvent losses can only be obtained by having a sufficiently high freeboard, preferably refrigerated.

Frit: a vitreous enamel with suitable electrical characteristics in powdered form and held in an organic binder system, used in screening inks for thick-film circuits.

Fungicide: a product used to kill living fungi and molds.

Fungostat: a product used to prevent fungi and molds from finding a medium on which to thrive.

Fusion: the process of melting, for example, a metal.

Fusion fluid: a liquid, vapor, or gas held at a suitable temperature to promote fusion.

Galvanic corrosion: synonym for electrolytic corrosion.

Gel: A hydrophilic colloid, if sufficiently concentrated, will set to a jelly-like consistency. This is due to the coagulation of the solids to pseudo-fibers, which will imprison the liquid. Most such products are reversible between gels and hydrosols.

Glass: a supercooled liquid forming a noncrystalline ceramic solid, often transparent, usually brittle, breaking with a conchoidal fracture. Silica, lime, and sodium carbonate, the ingredients for common "bottle" glass, produce a mixture of sodium and calcium silicates. Thousands of glass formulations exist, but only a few are used in electronics as substrates, reinforcements, vitreous enamels, thick film frits, etc., as well as optoelectronic components.

Glass electrode: sensitive element for the electrical determination of pH.

Glass transition temperature: Often abbreviated T_g, this is the temperature above which a synthetic resin will no longer break with a conchoidal fracture but will yield under stress. It is not constant for any one resin, varying according to the way in which the stress is applied, humidity content, and so on. That there is a gradual phase change around the T_g can be readily shown by measuring the thermal coefficient of linear expansion.

Glycol: any alcohol with two OH groups. Certain glycols have a tendency to polymerize, attack cross-linked polymers, or attach themselves by adsorption bonds to polymeric substrates. As their removal may be very problematic and they are all more or less hygroscopic, their use in flux compositions should be tempered with caution, particularly the heavier types. Glycols, their esters and ethers, are frequently used in the composition of water-soluble fluxes, some rosin fluxes, solder pastes, fusing and hot air leveling fluxes, water-soluble soldering oils, and fusing fluids. Their use is therefore a real problem in modern high-reliability electronics.

Gum: the natural exudation from the bark of certain trees, often polymeric if allowed to solidify.

Halide: an ion derived from a halogen.

Halogenated solvents: organic solvents in which at least one hydrogen atom has been replaced by a halogen atom.

Halogens: the elements fluorine, chlorine, bromine, iodine, and astatine. They are found in group 7a (p^5) of the periodic table.

Hardener: chemical used to promote cross-linking of linear polymers.

Hardness: a measure of calcium and magnesium salts in tap water. Various non-interchangeable methods are used for expressing the value, none of which is really useful without a more complete analysis.

Heat-transfer paste: temperature-stable silicone grease or other vehicle loaded with an electrically insulating, thermally conducting substance, such as beryllium oxide or alumina; used to ensure minimal thermal resistance between a semiconductor housing and a radiator. It should be used sparingly, as an excess may cause contamination.

Hybrid circuit: a form of printed circuit on an inorganic substrate, usually with printed components.

Hydrocarbons: organic compounds exclusively composed of hydrogen and carbon.

Hydrogen ion concentration: *see* pH.

Hydrolysis: a chemical reaction in which water is split into hydrogen and hydroxyl ions, which then react with weak acids or bases to dissolve them. All hydrolysis reactions can be catalyzed by the presence of strong acids, alkalis, or ether.

Hydrophilic: attracted to water.

Hydrophobic: repelled by water.

Hydrosol: a colloidal dispersion of microparticles in a liquid.

Hydroxyl: an anion consisting of one hydrogen and one oxygen atom $(OH)^-$.

Hygroscopic: able to absorb humidity from the atmosphere, without that substance necessarily being water soluble, *See* Deliquescence.

Inclusion: essentially, a particulate contaminant within the mass of a laminate. By extension, it also refers to nonparticulate contaminants and particulate contaminants encapsulated on the surface of a laminate or elsewhere by a conformal coating or other protection.

Intermetallic: a chemical compound with a stoichiometric composition, formed from two or more metallic elements.

Ion: an atom or molecule that has acquired an electrical charge by gaining or losing one or more valence electrons. For example, a molecule of salt, NaCl, can become ionized in a water solution by dissociation into a positive sodium ion, Na^+, and a negative chloride ion, Cl^-.

Ion exchange resins: plastic beads that are designed to capture cations, replacing them with hydrogen ions, and anions, replacing them with hydroxyl ions, or a mixture of both types. Used for purifying water and some polar solvents.

Ionic contamination: a contamination consisting of dissociable molecules and ionogens.

Ionogen: a substance that is nonionic in itself, but becomes ionic after hydrolysis. For example, glacial acetic acid is a nonconductor, hence nonionic. Mixed with water, it hydrolyzes to become ionic and dissociates: $CH_3COOH + H_2O = CH_3COO^- + H_3O^-$.

Ionograph: a registered trademark of Alpha Metals, Inc. for their instruments that measure ionic contamination.

Ionophore: a synonym for an ionic substance or one that can only exist by itself in the form of crystals within an ionic lattice. Salt (sodium chloride) is an ionophore.

Isomer: one of two or more molecules with the same molecular formula, atomic composition, and molecular weight, but different physical structure. For example, C_4H_{10} can exist as butane or as 2-methyl propane.

Kauri-Butanol value: an artificial index of the solvency of organic liquids, of little value in comparing the efficacy of solvents for flux removal.

Ketone: an organic compound containing a CO group attached to two other carbon atoms.

Kiss cleaning: flux removal (not decontamination) on the solder side only of a printed circuit board by solvent brushing. Used principally as a low-cost method allowing subsequent use of "bed of nails" in automatic testing equipment.

Laser: a visible or invisible source of coherent light, an acronym for Light Amplification by Stimulated Emission of Radiation; used as highly controllable energy sources for many purposes, such as soldering.

Leaching: the solvent extraction of a soluble substance from an insoluble one, characterized by a lengthy processing time to complete dissolution.

Lead frame: a metallic framework used for supporting the chip of an integrated circuit, bonding the various connections to it, and forming the housing around it. At the end of the process, the supporting framework is cut off to provide the tinned soldering leads.

Leakage current: an electric current passing between two conductors, across which there is a potential difference, but which should be perfectly insulated from one another. In the context of this document, this usually refers to a current passing between adjacent conductors on a circuit, along or close to the surface of the insulating substrate.

Lipophilic: attracted to grease.

Lipophobic: repelled by grease.

Liquidus: the temperature at which a noneutectic mixture becomes entirely liquid. The liquidus curve is plotted from the liquidus temperature against composition. The liquidus is the same as the solidus at the eutectic point.

Maximum allowable concentration (MAC): the peak concentration of vapors of chemicals permitted in a working environment, based on a worker absorbing them over an eight hour day and a five-day week, expressed in ppm. *See* Threshold Limit Values.

Mealing: *see* Vesication.

Measling: defect that may be apparent in a laminate where a whitish area appears at the weave crossover. One cause may be poor pressing, leaving little or no resin between the surface and the cloth (*see* Weave exposure). Another cause is unsatisfactory wetting of the cloth by the impregnating resin (*see* Treatment, glass), leaving minute air bubbles trapped in the crossover. Measled laminate is of doubtful quality.

Methylated spirits: a form of denatured ethanol designed as a domestic fuel for spirit lamps. It is so impure that it should never be used for flux thinning or removal.

Micelle: a group of long molecules with hydrophobic ends clumped together in water to form a pseudo-colloid.

Mixed bed resins: cationic and anionic ion exchange resins mixed in a single column to eliminate both ion types in one operation.

Modified rosins: natural rosins may be modified chemically to enhance certain properties in fluxes, even rendering them water-soluble.

Mole: the amount of a substance that contains the same number of elementary particles as there are atoms in 12g of carbon 12. The type of elementary particle should be specified.

Molecular formula: the chemical formula of a compound expressed in an empirical form expanded to the full molecular weight.

Molecular sieve: a particular form of mineral aluminosilicates (zeolites) with a porous structure; pore size is selected to absorb molecules of a specific size range. The most popular ones are used for cation exchangers in water softeners and for absorbing water out of hydrophobic solvents.

Molecular weight: the sum of atomic weights in a molecule.

Molecule: the smallest part of a chemical element or compound that can exist by itself in a free state.

Monolayer: a surface layer of oriented molecules one molecule thick. In some cases, van der Waals' forces can form extremely dense monolayers, which could be hygroscopic and difficult to eliminate.

Multilayer: a printed circuit with conductor patterns inside the substrate as well as outside, made by pressing together various single and double-sided circuits between which are sandwiched sheets of prepreg.

Neutralization: reaction whereby the hydrogen ion concentration is made equal to the hydroxyl ion concentration to form a salt and water.

Nonionic contamination: contamination which, if immersed in pure water, does not alter its conductivity.

OA flux: a water-soluble flux strongly activated with organic acids, amines, or hydrochlorides.

Omega Meter: a registered trademark of Kenco Industries, Inc. for their instruments for measuring ionic contamination.

Organometallic salt: any salt whose anion is of organic origin, obtained typically as a reaction product between a metal or its oxide and an organic acid.

Osmosis: the passage of water through a semi-permeable membrane into a solution.

Osmotic pressure: the pressure needed to apply to a solution under osmosis to prevent water from flowing in either direction.

Outgassing: common phenomenon producing blowholes and craters in solder joints, usually caused by some form of contamination in or behind the barrel of a plated through-hole being released by applied heat.

Overhang: a dangerous source of particulate contamination caused by etching under the edge of a metallic resist. Overhang is prone to break off under mechanical stress and may cause short circuits. With tin-lead plated circuits, overhang is eliminated by fusion of the plating. Any ionic contamination, like etch residues, trapped under an overhang may provoke dangerous corrosion, especially if the conductor is gold-plated.

Oxide: compound of any element with oxygen. In this context, metallic oxides are common contaminants, especially as a cause for lowering the solderability of the metals.

Pads: areas on a printed circuit where connections are to be made.

Particulate contamination: contamination by physical particles.

Perhalogenated, perchlorinated, or perfluorinated solvents: halogenated solvents in which all the hydrogen atoms have been replaced by one type of halogen atoms. For example, carbon tetrachloride, CCL_4, is a perchlorinated solvent.

Permanent hardness: calcium or magnesium salts in tap water that do not precipitate out upon heating.

Persulfates: salts of persulfuric acid or peroxidisulfuric acid ($H_2S_7O_8$), usually the ammonium or sodium salts; very strong oxidizing agents but with a slow, easily controllable reaction. They are useful for oxidic cleaning of metals, notably copper, with a light etch, and are often incorporated in other detersive and chelating products for promoting highly reactive metallic surfaces. Ammonium persulfate produces insoluble complexes when attacking copper and a post-cleaning stage of dilute sulfuric acid is essential.

pH: an artificial measure of the acidity or alkalinity of an aqueous solution. It is the negative logarithm to the base of 10 of the hydrogen ion concentration expressed as gram-ionic weights per liter. Pure water has an approximate hydrogen ion concentration of 10^{-7} moles per liter at 24°C, so it may be considered as having a neutral pH of 7. Acids have lower and bases have higher values of pH.

Phase: a homogeneous state of matter (solid liquid, gas).

Phase diagram: a graphic representation showing the equilibrium conditions between the various phases of a substance, relating to temperature, composition, crystallization, pressure, and so on.

Phenol: any compound with a hydroxyl group attached to any carbon atom in an aromatic ring.

Phenolic resins: a series of synthetic resins formed by reacting phenols with aldehydes.

Photoresist: a liquid or dry film resist that can be developed after exposure.

Plasma cleaning: a method of organic contamination removal by placing the part or assembly in a plasma of a carefully selected gas. The same technology is used for production desmearing of holes prior to through-hole plating, especially for multilayer circuits.

Plastic: a generic term in common usage covering a wide range of synthetic polymers.

Plasticizers: products added to polymers to render them more flexible. Many plastic films contain plasticizers, which may be contaminants for electronic assemblies.

Plate: the component of a reflux distillation unit designed to ensure optimum contact between the ascending vapor and the descending reflux.

Polar solvent: a solvent whose molecules possess an unbalanced electrical charge due to an asymmetrical structure. The polarity is determined by the vector sum of the individual positive and negative charges, producing a dipole the field strength of which is inversely proportional to the square of its vector length. The most strongly polar of the common solvents is water, with methanol, ethanol, and the other alcohols in descending order; water has a field strength about fourteen times greater than the simplest organic polar solvents, which is why it is so effective for removing ionic contaminants.

Polycarbonate: a transparent polymer with good electrical and mechanical properties, but chemically rather tender, used for some components.

Polyester: a polymer produced by the condensation reaction of the esterization of polyalcohols and polybasic acids.

Polyethylene: a simple polymer with excellent electrical properties and good chemical resistance, but poor mechanical properties. Its use as flexible film plastic bags should be tempered with caution as it tends to be porous and contains often undesirable plasticizers.

Polyglycol: *see* Glycol.

Polyimides: a group of synthetic resins where cross-linking is achieved through the groups O-C-NH-C-O or O-C-N-C-O or their close relatives. Polyimide resins are more thermally stable than epoxy resins, but are electrically poorer owing to greater water absorption.

Polymer: a product that has undergone polymerization. All "plastics" are polymers, as are many other natural and synthetic substances. Polymers can exist in any state of matter.

Polymerization: grouping molecules, or parts of molecules, into linear chains without modifying their chemical composition.

Polyolefines: a group of polymerized alkenes, of which the best known is polypropylene, an excellent, light, constructional thermoplastic also used as wrapping film. Certain qualities are noncontaminating, plasticizers being unnecessary.

Polystyrene: a polymer that is a superb high-frequency electrical insulator, but mechanically and chemically rather poor. It is used for low-loss components and for special lacquers. Great care should be taken not to damage it by the use of even military-quality solvents.

Polytetrafluoroethylene: usually abbreviated to PTFE. A perfluorinated version of polyethylene. PTFE has some excellent electrical and thermal properties, but mechanically is difficult to use, since it cannot be molded or transformed, only sintered. Its nonstick properties make it ideal for low "stiction" applications, but render it difficult to bond to other surfaces. A special treatment with sodium solutions can improve the bond strength, but it is never as good as with other polymers.

Polyurethanes: a group of synthetic resins formed by the condensation polymerization of polyhydroxy compounds with diisocyanates. The monomers and thermal decomposition products are highly toxic.

Polyvinyl chloride: or PVC, a rigid thermoplastic with excellent constructional qualities. It should be used with moderation, especially when temperatures higher than 50°C may be encountered, as it is easily decomposed, releasing monovinyl chloride (highly toxic) and hydrochloric acid. In a plasticized form, it is the most popular insulation for wires and cables, but is often avoided in high-reliability electronics, in favor of more sophisticated and stable polymers.

Prepreg: a common term for B-stage resin impregnating glass cloth. A dried and partially polymerized epoxy resin cloth.

Printed circuit: component consisting of an organic insulating substrate with a metallic conductor pattern.

Printed circuit board: assembled and soldered printed circuit, ready for use.

Pyrolysis: chemical decomposition due to the effect of heat; more specifically, the result of heat on a flux or polymer, especially due to a soldering or desoldering operation.

Quaternary ammonium hydrohalides: the product of the reaction between tertiary amines and alkyl halides, by which an ammonium ion (NH_4^+) has its four hydrogen atoms replaced by simple or aliphatic hydrocarbon chains, at least one of which ends in a hydrohalide group. The hydrochlorides and hydrobromides are used extensively as highly efficient rosin flux activators.

R flux: a rosin flux, nonactivated.

RA flux: a rosin flux, activated with organic hydrochlorides.

Reflux distillation: a form of distillation in which a dephlegmator condenses a proportion of the liquid and sends it in a counterflow through the vapor to scrub it, making it progressively cleaner.

Relative humidity: the ratio, expressed as a percentage, of the weight of water vapor in a given quantity of air to that which would saturate the same quantity of air at the same temperature and pressure.

Release agents: products used to prevent polymers from undesirably adhering to parts. In laminate and multilayer production, the release agent used is commonly a modified polyolefine or fluorinated plastic film. For molded components, an aqueous colloid may be used. On no account should silicone release agents be used for any application in the electronics industry. Release agents are also called adherents.

Reliability: a philosophical concept that has been quantified into mathematical relations. The overall reliability of an electronics assembly is dependent on numerous factors, according to the application. Contamination control is one such factor that is applied to all high-reliability electronics; it is sometimes applied to other professional electronics, but rarely to other applications. The current tendency is for contamination control to be increasingly applied to less critical applications to reduce cost, as well as to increase reliability.

Resin: any substance of a variable form, softening at high temperatures, of vegetable, animal, or synthetic origin. Resins are polymeric. With crosslinked thermosetting resins, softening may not occur before pyrolysis. Rosin and resin are also sometimes used as synonyms.

Resin flux: a flux using a modified rosin or a synthetic resin.

Reverse osmosis: the passage of water out of a solution through a semi-permeable membrane, promoted by the application of high pressure to the solution. It is not a filtration.

Reversion: the partial or complete depolymerization of a polymer back into its original monomers. In the context of the electronics industry, it is usually used to describe a partial depolymerization, often due to pyrolysis. This phenomenon is exploited in polyurethane solderable enamels used for insulating magnet winding wires. Of the plastics commonly used in the electronics industry, the polyurethanes are the most apt to revert.

Rheology: the manner in which a fluid flows. If it is a perfect fluid, the rheology will be an exact function of its viscosity.

RMA flux: a rosin flux, mildly activated with organic acids. RMA fluxes should not have free halides in their composition.

Rosin: a resin mixture, mainly of isomers of abietic acid, obtained from diverse pine trees and residual after distilling of the extract. There are two methods of extraction, gum from the living tree and chips of a felled tree. As a flux, gum rosin is considered superior, although a good wood rosin is not much different.

Rosin flux: a flux whose principal active component is rosin, either activated or not.

Rust: hydrated iron oxide on the surface of iron or steel, produced as an autoelectrolytic corrosion reaction. In addition to the metal, an included impurity (iron carbide or carbon), water, and ions are necessary to complete the electrolytic couple.

SA flux: a solvent-soluble flux, strongly activated with organic nitrates and phosphates.

Salicylic acid: a monocarboxylic aromatic acid used as a rosin flux activator. Decomposes entirely to volatile components at soldering temperatures, but the physiological properties of both the acid (closely related to aspirin) and its decomposition products (phenol, etc.) render its use hazardous in significant quantities.

Salt: any substance that can yield cations (other than hydrogen ones) and anions (other than hydroxyl ones). The reaction product between an acid and a metal or its oxide.

Saponification: the hydrolysis of an ester to a soap via a caustic alkali, with glycerol as a by-product; by extension, the reaction between any insoluble carboxylic acid and an alkali. Widely used in the electronics industry to render rosin flux residues water-soluble by reacting them with aqueous solutions of organic amines.

Self-ionization: partial dissociation of water or other solvents: $2H_2O = (H_3O)^+ + OH^-$.

Silane: literally, a silicon hydride in the family from SiH_4 to Si_8H_{18}. The term is also used for many products with a hydrolysable group at one end of a polymer, where the end carbon atom is replaced by a silicon atom and the other end is hydrophobic. The hydrophilic end can be made to wet glass and the other end can crosslink into an epoxy resin. These complex products are used for treating glass cloths for laminate manufacture to promote optimum adhesion between the glass fibers and the epoxy resin.

Silica gel: a desiccant produced by dehydrating of a gel formed by the hydration of amorphous silicon dioxide. It absorbs water without reverting to its gel state. It can be regenerated by heat but only a limited number of times, as the absorbent zones tend to become blocked by impurities in the absorbed water.

Silicones: organic compounds in which one or more carbon atoms have been replaced by silicon atoms, generally useful in polymeric forms (greases, varnishes, resins, rubbers, etc.). Should be employed with circumspection in the electronics industry.

Silk screening: printing process by which ink is forced through a cloth-bound stencil onto the object being printed by means of a elastomeric blade.

Sintering: fusion of powders into masses at temperatures below their melting point, usually by the application of high pressure.

Smearing: the result of a mechanical operation in which a blunt cutting tool, or similar object, heats to a temperature above the softening point of the material being worked, then smears it over a badly worked surface. Can be catastrophic in drilling certain types of printed circuits.

Soap: an organometallic salt with micelle-forming properties, used for detersive cleaning. Commercial soaps are generally made by saponifying vegetable and animal fats with sodium hydroxide, producing soap and glycerol. The latter is subsequently removed.

Solder: an alloy used for joining metals of a higher melting point, able to form intermetallics with them to provide bonding.

Solder balls: a common particulate contamination resulting from the use of solder pastes under less-than-ideal conditions or from spattering caused by wet fluxes during wave soldering.

Solder cream: *see* Solder paste.

Soldering oil: a mineral or water-soluble product designed to reduce the oxidation on a molten solder surface and to render it more fluid by reducing surface tension.

Solder paste: a suspension of solder powder in a flux base, with additives to confer suitable rheological properties. The flux residues after soldering are notoriously severe contamination.

Solder resist or mask: interchangeable terms to denote a permanent polymeric coating applied to the whole surface of a printed circuit, except to areas where soldering is subsequently required.

Solder stop: a temporary solder resist that can be removed after soldering, either by peeling it or dissolving it in the solvent used for cleaning.

Solid solution: a colloidal dispersion of microparticles in a solid.

Solidus: the temperature below which a noneutectic mixture becomes entirely solid. The solidus is the same as the liquidus at the eutectic point, and frequently is the same temperature over a range to either side of it.

Solubility: an expression of the maximum weight of a solute that a given quantity of solvent can hold in solution at a given temperature.

Solubilization: means by which an insoluble product may be rendered soluble, e.g., using a detergent to render oil soluble in water.

Solute: a reactive part of a solution. Contamination becomes a solute after having been cleaned off by a solvent.

Solution: the result of a solute dissolved in a solvent.

Solvent: a nonreactive part of a solution.

Stearic acid: monobasic carboxylic acid used as a rosin flux activator.

Stress corrosion: metallic corrosion enhanced by the presence of stresses in the metal.

Stripping: in this context, the removal of a temporary mask (photoresist, plating resist, etc.) by chemical means. Either strong alkalis or strong solvent mixtures are generally used for stripping.

Structural formula: a schematic representation of a chemical formula in a bidimensional or tridimensional format showing the relative positions of the atoms.

Substrate: in this context, the insulating support of printed or hybrid circuits.

Succinic acid: dibasic carboxylic acid used as a rosin flux activator, but irritating to the mucous membranes.

Sulfonic acids: organic acids with a mild chelating surfactant action, often used as copper cleaners.

Surface active agents: long-chain detergents designed to reduce the surface tension of water at low concentrations.

Surface-mounted components: components soldered directly onto the surface of a printed circuit, without wires or leads passing through the holes.

Surface tension: the physical effect apparent at gas-liquid, gas-solid, liquid-liquid, and liquid-solid interfaces; governed by a series of apparent tensions in the fluids.

Surfactants: surface active agents that form oriented monolayers at liquid-gas interfaces.

Tall oil: crude product, before distillation and purification, obtained from pine wood chips, from which wood rosin can be obtained.

TDI: *see* Toluene diisocyanate.

Temporary hardness: calcium or magnesium salts in tap water that precipitate out on heating.

Terpenes: group of aromatic hydrocarbon isomers with the formula $C_{10}H_{16}$, with boiling points of about 180°C. They are found naturally in many plants, including spices, citrus fruits, herbs, and conifers.

Thermoplastics: polymers, mainly with a linear structure, whose composition does not alter when taken up and down repeatedly through their melting points, which are reasonably clearly defined. PTFE is an exceptional thermoplastic, in that the melting point is undefined.

Thermosetting resin: a polymer that can be hardened into a permanent form by the application of heat and, once hardened, cannot be made to melt again, e.g., epoxy resin.

Thick-film circuit: a hybrid circuit, usually on a ceramic or glass substrate, with conductor, resistor, and insulator patterns silk-screened successively, using frits, and fired in a furnace.

Thin-film circuits: a hybrid circuit, usually on a ceramic or glass substrate, with conductor, resistor, and insulator patterns successively evaporated through masks.

Thixotropy: the property of a fluid that lowers its viscosity as the shear forces within it increase, such as may be exhibited by a gel being forced into a hydrosol.

Threshold limit values (TLV): time-weighted average concentrations of vapors of chemicals permitted in a working environment, based on a worker absorbing them over an eight-hour day and a five-day week, expressed in ppm.

Toluene diisocyanate: extremely toxic and irritating component of some synthetic resins, notably polyurethanes.

Transducer: an electromechanical device working either on the piezoelectric or magnetostriction principle for inducing ultrasonic vibrations in a liquid.

Also a general term for any device that converts one form of energy into another.

Treatment (copper): any one of a number of different processes to promote the adsorption of epoxy resins by copper in the manufacture of laminates (peel strength). Typical treatments are chemical oxidation (blackening) and the electrochemical deposit of zinc (which will alloy with the copper at pressing temperature), or of other alloys, usually with a structure to provide as large a surface area as possible. The treatment should be such that residues are removed during etching, otherwise contamination will result. Mechanically, parts of the treatment should not break off during pressing, or the electrical qualities of the laminate will suffer.

Treatment (glass): the adhesion of epoxy resins to pure glass is not very good unless the glass is treated specially to promote it. The most useful treatments are with silanes, but organometallic chromium salts may also be used for the same purpose. The latter may be more dangerous in the case of weave exposure.

Turpentine: nonscientific name of the distillate obtained from the extraction of gum or wood rosin, consisting mainly of a mixture of terpenes.

Ultrasonic cleaning: cleaning in water, organic solvents, or solutions with a mechanical ultrasonic energy applied at a frequency and amplitude to create cavitation. The implosion of the cavities creates an adiabatic compression, producing an intense energy level over a minute distance.

Van der Waals' forces: adsorption forces enabling molecules of a micelle-forming type to pack on a surface in densities many times the normal ones, overcoming the natural repulsion forces of weak dipoles aligned side-by-side.

Vapor: a gaseous phase of a liquid.

Vapor degradation: the accumulation of impurities in the vapor phase of a solvent cleaning machine.

Vapor phase cleaning: a method of cleaning by which a clean solvent vapor is condensed on a cold object and allowed to run off until the object attains the temperature of the vapor. Used mainly as a final stage after immersion or spray cleaning in organic solvents.

Vapor phase soldering: a method of reflowing a solder paste, preform, or tinned surface by allowing a solvent to vapor, i.e., at some 30 to 35°C higher than the solder melting point to condense it anaerobically.

Vesication (mealing): the formation of minute blisters (vesicles) under a coating due to the absorption of water vapor by an imprisoned contaminant.

Viscosity: resistance of a fluid to a change in form imposed by the application of an external force. Various ways of measuring it give somewhat different results. It can only be expressed in units within the context of a given set of measurement conditions, unless it is a perfect liquid.

Water softener: a device for replacing calcium and magnesium ions in hard water by sodium ones.

Wave soldering: a method of mass soldering components on a printed circuit by passing the prefluxed assembly over a pumped molten solder wave under controlled conditions. For surface-mount components, a double or hollow wave system is sometimes used.

Weave exposure (measling): defect when there is no resin between the outer surface of a laminate and the weave crossover. It may be due to poor pressing, a chemical attack on the surface, or, more likely, a combination of both causes.

Whisker: a metallic growth, from several different causes, able to form short circuits. Tin and silver are two metals particularly prone to whisker formation.

White deposit: any one of a number of effects that appear white after flux residue removal.

Wood rosin: obtained from a felled tree, as opposed to gum rosin, which is obtained from tapping the living tree.

Zeotrope: a mixture of liquids whose boiling point is not constant and/or whose composition in vapor phase is different from that of the liquid phase.

REFERENCES

[1]   Grayson, M., *Encyclopedia of composite materials and components*, John Willey & Sons: New York, 1983. pp. 193.

[2]   Turlik, I., *Advanced packaging technology*. Microelectronics Center of North Carolina: North Carolina, 1994.

[3]   *IPC Standard EG-140 Specification for Finished Fabric Woven E-glass for Printed Boards*. The Institute for Interconnecting and Packaging Electronic Circuits: Northbrook, IL, March 1988.

[4]   Tautscher, C. J. *Contamination effects on electronic products*. Marcel Dekker, Inc.: New York, 1991.

[5]   Mayo, R., Nanni, A., Watkins, S., "Strengthening of Bridge G-270 with externally bonded carbon fiber reinforced polymer (CFRP)," Research Investigation RI98-02, Missouri Department of Transportation, December 1999. p. 4.

[6]   *The Condensed Chemical Dictionary*, 8th Edition. Hawley G., revised. Van Nostrand Reinhold Company: New York. 1971.

[7]   Fullwood, L. "Removing negative acting photoresists," *Electronic Packaging and Production*. December 1971.

[8]   Tautscher, C., *The contamination of printed wiring boards and assemblies*. Omega Science Services: Washington, 1976.

[9]   "Adhesives for composite materials used in printed circuitry." U.S. Department of Commerce, OTS, PB 121960.

[10]   Zado, F., "Insulation resistance degradation caused by nonionic water soluble flux residues," *Proceedings of the Technical Program Nepcon*, 1979. p. 374-376.

[11]   Ellis, B. N., "New aspects on the solderability of printed circuits," *Proceedings of the Technical Program, Nepcon*, 1969.

[12]   Boudreau, R. J., "Glass fabric finishes: Effects on the kinetics of water, absorption, and laminate physical and electrical properties," *Proceedings of the Printed Circuit World Convention II, 1*. 1981.

[13]   Ellis, B. N., "Contamination on components and small assemblies," *Proceedings of the Technical Program, Internepcon*. 1984.

[14]   Lea, C., Howie, F., H., "Controlling the quality of soldering of PTH solder joints." *Circuit World*, Vol.11, No.3, 1985. pp. 5-7.

[15]   Matthaes, G., "The technique of surface cleaning of professional printed circuit boards and multilayers," *Technical Data Sheet*. 1981.

[16]   Toms, K. R., "…but what are we doing to the copper?" *Circuit World*, Vol. 8, No. 1. 1981. p. 24.

[17]   ASM Handbook, Volume 5, Surface Engineering. *Choosing a Cleaning Process*. ASM International. Materials Park. 1996.

[18]   ASM Handbook, Volume 5, Surface Engineering. *Guide to Mechanical Cleaning Systems*. ASM International. Materials Park. 1996.

[19]   ASM Handbook, Volume 5, Surface Engineering. *Guide to Alkaline, Emulsion, and Ultrasonic Cleaning*. ASM International. Materials Park. 1996.

[20]   Crawford, T., "Conformal coating adhesion over no-clean fluxes," *Annual Electronics Manufacturing Seminar*, 1995. pp. 21-30.

[21]   Ritchie, B., Conformal coating over soldermask, no-clean soldering and alternate cleaning technologies. *Proceedings of the Technical Program, NEPCON West '94*. Vol. 2, 1994. pp. 1312-1331.

[22]   Christou, A. *Electromigration and electronic device degradation*. John Wiley & Sons: New York, 1994.

[23]   Scully, J. C. *The fundamentals of corrosion*. Pergamon Press: Oxford, 1990.

[24]   Bennington, L. D., Crossan, D., "Conformal coating overview," *Proceedings of the Technical Program, NEPCON West '95*, Vol. 3, 1995. pp. 1757-1761.

[25]   Cantor, S. "Solvent-free aerobic adhesives and coating III novel multi-cure conformal coating provides complete shadow cure at room temperature," *Proceedings of the Technical Program Nepcon West '96*, 1996. pp. 443-457.

[26]   Crain, K. "Conformal coatings in microns," *Surface Mount Technology*, Vol. 9, 1995. pp. 50-52.

[27]   Crum, S. "Solvent-free conformal coatings impact application techniques," *Electronic Packaging and Production*, Vol. 37, June 1997. pp. 61-70.

[28]   Cummings, R. J., "Conformal coatings: An overview of processes and removal methods," http://arioch.gsfc.gov/eee_links/vol_02/no_03/eee2-3r.htm, as viewed on July 8 1998.

[29]   Frost, G. T., "Making a choice: no VOC/low VOC [PCB conformal coatings]," *Proceedings of the Technical Program, Nepcon West '95*, Vol. 3, 1995. pp. 1745-1748.

[30]   Horrocks, H., "Conformal coating removal - what is the best method," *Proceedings of the Technical Program, Nepcon West '96*, 1996. pp. 1064-1075.

[31]   Martin, V. D., "Stripping conformal coatings [PCB repair]," *Electronic Packaging and Production*, Vol. 35, 1996. pp. 51-52.

[32]   Naisbitt, G., "Conformal coating coverage." *TECHNET news group posting*. September 25 1997.

[33]   Olson, R. "High performance, non-ODC/VOC conformal coating," *Proceedings of the Technical Program, NEPCON West '95*, Vol. 3, 1995. pp. 1762-1767.

[34]   Olson, R., "Vacuum-deposited conformal coatings...PCB protection option," *Surface Mount Technology*, Vol. 5, 1991. pp. 37-43.

[35]   Quade, R., "Conformal coatings and process options," *NEPCON West '95*, Vol. 3, 1995. pp. 1749-1756.

[36]   Rybczyk, J., "Conformal coating process improvement," *Circuits Assembly*, Vol. 7, No. 8, 1996. pp. 44, 46, 48, 50, 52.

[37]   Studd, R., "A conformal coating evol.ution," *Surface Mount Technology*, Vol. 11, No. 4, April 1997. pp. 48-50.

[38]   Tautscher, C. J., "Conformal coatings selection criteria," *Surface Mount Technology*, Vol. 12, No. 7, July 1998. p. 64.

[39]   Tautscher, C. J., "A focus on conformal coating," *Circuits Assembly*, Vol. 8, No. 5, 1996. pp. 30-2, 34, 37.

[40]   Brous, J., "Cleaning materials," *Soldering and Mounting Technology*. Alpha Metals, Inc. 1996.

[41]   Hymes, L., *Cleaning printed wiring assemblies in today's environment*. Van Nostrand Reinhold: New York, 1991.

[42]   Constance, J. D., *Controlling in-plant airborne contaminants*. Marcel Dekker, Inc.: New York, 1983.

[43]   Clark, R. H. *Printed circuit engineering*. Van Nostrand Reinhold: New York, 1989.

[44]   Bogenschütz, A. F., *Analysis and testing in production of circuit boards and plated plastics*. Finishing Publications, Ltd.: Teddington, 1985.

[45]   Shineldecker, C. L., *Handbook of environmental contaminants: a guide for site assessment*. Lewis Publishers: Boca Raton, 1992.

[46]   Prasad, R. P., *Surface mount technology: principles and practice*. Van Nostrand Reinhold: New York, 1989.

[47]   Manko, H. H., *Soldering handbook for printed circuits and surface mounting*. Van Nostrand Reinhold: New York, 1995.

[48]   Geragosian, G., *Printed circuit fundamentals*. Reston Publishing Company: Reston, 1985.

[49]   Boswell, D., Wickham,M., *Surface Mount Guidelines for Process Control, Quality, and Reliability*. McGraw-Hill: London, 1992.

[50]   Matthew, L. C., Rath, D. L., *The waterdrop test highly accelerated migration testing*. IBM T.J. Watson Research Center: New York, 1991.

[51]   Tegehall, P. E., "Reliability verification of printed board assemblies: a critical review of test methods and future test strategy," *Surface Mount International Exposition*. August, 1998. pp.359-382.

[52]   Kenyon, W. G., "Ionic testing: questions and concerns," *Surface Mount Technology*, Vol. 12, No. 5, May 1998.

[53]   Kenyon, W. G., "Borrowing from the semiconductor industry," *Surface Mount Technology*, Vol. 11, No. 10, October 1997.

[54]   Munson, T., "Understanding residues," *Surface Mount Technology*, Vol. 11, No. 10, October 1997. pp. 66.

[55]   Tautscher, C. J., "The evolution of electronic cleaning," *Circuits Assembly*. Vol. 9, No. 3, March 1998. pp. 74, 76, 78.

[56]   Benedict, H., Yuen, M., "A batch cleaner optimization study," *Circuits Assembly*. Vol. 9, No. 10, October 1998. pp. 52, 54-7, 59.

[57]   Finch, K., "Comparison of synthetic semi-aqueous alcohol cleaning technology," *Surface Mount Technology*, Vol. 13, No. 2, February 1999. p. 82.

[58]   Kelly, L., "Removing contamination, improving yield the just-in-time solution," *Electronic Packaging and Production*, Vol. 38, No. 15, December 1998. pp. 38-42.

[59]   Markstein, H. W., "Cleaning assembled circuit boards," *Electronic Packaging and Production*, Vol. 38, No. 12, October 1998. pp. 26-34.

[60]   McCall, P. "No-clean no hazard?" *Surface Mount Technology*. Vol. 12, No. 8, October 1998. pp. 68-71.

[61]   Crawford, T. "Cleaning," *Special Supplement to SMT: Back to Basics*. July 1998.

[62]   Kohman, G., Hermance, H., Downes, G., "Silver migration in electrical insulation," *The Bell System Technical Journal*, 1955. pp. 1115-1147.

[63]   Lando, D., Mitchell, J., Welsher, T., "Conductive anodic filament in reinforced polymeric dielectrics: formation and prevention," *Proceedings of the Seventeenth International Reliability Physics Symposium*, 1979. pp. 51-63.

[64]   Pecht, M., Wu B., Jennings, D., "Conductive filament formation in printed wiring boards," *Thirteenth IEEE/CHMT International Electronics Manufacturing Technology Symposium*, 1992. pp. 74-79.

[65]   Lahti, J., Delaney, R., Hines, J., "The characteristic wearout process in epoxy-glass printed circuits for high density electronic packaging," *Proceedings of the Seventeenth International Reliability Physics Symposium*, 1979. pp. 39-43.

[66]   Krumbein, S., "Metallic electromigration phenomena." *IEEE Transactions on Components, Hybrids, and Manufacturing Technology*, Vol. 11, No. 1, 1988. pp. 5-15.

[67]   Rudra, B., Pecht, M., Jennings, D., "Assessing time-to-failure due to conductive filament formation in multi-layer organic laminates," *IEEE Transactions on Components, Packaging, and Manufacturing Technology, Part B*, Vol. 17, No. 3, 1994. pp. 269-276.

[68]   Rudra B., Jennings, D., "Failure-mechanism models for conductive-filament formation," *IEEE Transactions on Reliability*, Vol. 43, No. 3, 1994. pp. 354-360.

[69]   Shukla, A., Pecht, M., Jordan, J., Rogers, K., Jennings, D., "Hollow fibers in PCB, MCM-L and PBGA laminates may induce reliability degradation," *Circuit World*, Vol. 23, No. 2, 1997. pp. 5-6.

[70]   Shukla, A., Dishongh, T., Pecht, M., Jennings, D., "Hollow fibers in woven laminates," *Printed Circuit Fabrication*, Vol. 20, No. 1, 1997. pp. 30-32.

[71]   Pecht, M., Hillman, C., Rogers, K., Jennings, D., "Conductive filament formation: A potential reliability issue in laminated printed circuit cards with hollow fibers," *IEEE Transactions on Electronics Packaging Manufacturing*, Vol. 22, No. 1, 1999. pp. 80-84.

[72]   ANSI/IPC-A-600 *Acceptability of printed boards*. The Institute for Interconnecting and Packaging Electronic Circuits: Northbrook, IL, August 1995.

[73]   *Acceptability of electronic assemblies: Revision B*. The Institute for Interconnecting and Packaging Electronic Circuits: Northbrook, IL, December1994.

[74]   *Post solder solvent cleaning handbook*. The Institute for Interconnecting and Packaging Electronic Circuits: Northbrook, IL, 1987.

[75]   *Post solder aqueous cleaning handbook*. The Institute for Interconnecting and Packaging Electronic Circuits: Northbrook, IL, 1996.

[76]   Ferlauto, E., "A study of the effect of acid rain on alkyd, polyester and silicone-modified high-solids coatings," *Federation of Societies for Coatings Technology*. October 1994.

[77]   Tegehall, P.E., "Evaluation of the capability of conformal coatings to percent degradation of printed circuit assemblies in harsh environments part one," *IVF Research Publication 93803*, 1993.

[78]   Tegehall, P. E., "Influence of application method on the reliability of conformal coated circuit assemblies," *IVF Research Publication 96809*, 1996.

[79]   Tegehall, P. E., "Evaluation of the effect of various combinations of solder masks and flux residues on the reliability of conformal coated circuit assemblies," *IVF Research Publication 96808*, 1996.

[80]   Tegehall, P. E., "Evaluation of the capability of conformal coatings to prevent degradation of printed circuit assemblies in harsh environments part two," *IVF Research Publication 96807*, 1996.

[81]   Tegehall, P. E., "Evaluation of the influence of flux residues on the reliability of conformally coated PCB assemblies," *IVF Research Publication 93811*, 1993.

[82]   Tegehall, P. E., "Evaluation of test methods and the impact of contamination from production processes on the reliability of printed circuit board assemblies," *IVF Research Publication 96846*, 1996.

[83]   Fukuda, Y., Fukushima, T., Sulaiman, A., Musalam, I., Yap, L. C., Chotimongkol, L., Judabong, S., Potianart, A. Keowkangwal, O., Yoshihara, K., Tosa, M., "Indoor corrosion of copper and silver exposed in Japan and ASEAN countries," *The Journal of the Electrochemical Society*, Vol. 138, No. 5, May 1991. pp. 1238-1242.

[84]   IPC (1989). PWB Design Workshop, Lincolnwood, IL.

[85]   White, D. F., Fornes, R. E., Gilbert, R. D., Speer, J. A., "Investigation of the formation of Physical Damage on Automotive Finishes Due to Acidic Reagent Exposure," *Journal of Applied Polymer Science*, Vol. 50, 1993. pp. 541-549.

[86]   Miller, V. M., "Characterization of component failure modes and associated causes," *Advances in Electronic Packaging-ASME*, November 1999. pp. 1341-1347.

[87]   Agnew, J., "Choose cleaning solvents carefully," *Electronic Design 5*, March 1974. pp. 54- 57.

[88]   Capillo, Carmen. *Surface Mount Technology*. McGraw-Hill: New York, 1990.

[89]   Cordes, F., Huemoeller, R., "Electroless Ni/Au plating capability study of BGA packages," *Future Circuits International*. Issue 4. pp. 131 –134.

[90]   Biunno, N., "A root cause failure mechanism for solder joint integrity of electronic Ni/Au surface finishes," *Future Circuits International*. Issue 5. pp. 133 –139.

[91]   Gudeczauskas, D., Hashimoto, S., Kiso, M., "Gold Plating," *PC Fabrication*. Vol 22, No. 7, July 1999. pp. 30-3.

[92]   Mei, A., Johnson, P., Eslambolchi, A., Kaufmann M., "The effect of
         electroless Ni/Au plating parameters on PBGA solder joint attachment
         reliability," Proceedings of the ECTC, 1998. pp. 952-961.

[93]   Zribi, A., Chromik, R.R., Presthus, R., Clum, J., Zavalig, L., Cotts, E.J.
         "Effect of Au-intermetallic compounds on mechanical reliability of Sn-
         Pb/Au-Ni-Cu joints," *Advances in Electronic Packaging.* Vol. 26-2, no.2,
         1999. pp. 1573-1577.

[94]   Hayes, Michael E., "Semi-Aqueous defluxing using closed-loop
         processes," Paper IPC-TP-898: Second International Conference on Flux
         Technology, 1990.

[95]   Lowell, Charles R., Sterritt, Janet R., "An aqueous cleaning alternative to
         CFCs for rosin flux removal," Paper IPC- TP-893: Second International
         Conference on Flux Technology, 1990.

[96]   Marczi, M., Bandyopadhay, N., Adams, S., "No-clean, no-residue
         soldering process," *Circuits Manufacturing*, Vol. 30, February 1990. pp.
         42-46.

[97]   Murray, J., "Ultrasonic cleaning: what's the buzz?" *Circuits
         Manufacturing*, Vol. 30, January 1990. pp. 72-74.

[98]   Parikh, G., "AT&T achieving environmental and quality goals," *Circuits
         Assembly*, Vol. 2, December 1991. pp. 60-62.

[99]   Pickering, R., "Digital's microdroplet aqueous cleaning project," Paper
         IPC- TP-82: Second International Conference on Flux Technology, 1990.

[100]   Digital Equipment Corporation. "Aqueous microdroplet module cleaning
         process." ICOLP, 1990.

[101]   Russo, John F., and Fischer, Martin S., "Recycling boosts aqueous
         cleaners," *Circuits Manufacturing*, Vol. 29, July 1989. pp. 34-39.

[102]   Sinclair, J. D., "The relevance of particle contamination to corrosion of
         electronics in processing and field environments", Bell Laboratories,
         Electrochemical Society Semiannual International Meeting, Honolulu, HI,
         Oct. 1999.

[103]   Specialty Coating Systems. "Parylene conformal coatings specifications
         and properties." http://www.scscookson.com/parylene/properties.htm, as
         viewed on February 8 2002.

[104]   Virmani, N., "Understanding the effectiveness of parylene coatings in the
         protection of plastic encapsulated microcircuits and assemblies from
         moisture initiated failure mechanisms,"
         http://epims.gsfc.nasa.gov/ctre/ct/techdocs/parylene/parylene.html, as
         viewed on 26 August 2001

[105]   Neale, R., "COTS and performance based specifications – the blind leading the blind?" *Electronic Engineering*, October 2001. pp. 17-20.

[106]   Ritchie, B. "New conformal coatings combine protection with environmental safety," *Surface Mount Technology*, April 2000. pp. 16-19.

[107]   *IPC-HDBK-830. Guidelines for Design, Selection, and Application of Conformal Coatings*. The Institute for Interconnecting and Packaging Electronic Circuits: Northbrook, IL, October 1999.

[108]   Licari, James J., Hughes, Laura A., *Handbook of Polymer Coatings for Electronics*. Second Edition, Noyes Publications: Park Ridge, New York, 1990.

[109]   Wong, C. P., Pike, Randy T., and Wu, J., "Interface-adhesion-enhanced bi-layer conformal coating for avionics application," *Proceedings of the International Symposium on Advanced Packaging Materials*, 1999. pp. 302-10.

[110]   Dietrich, Tim. "Deep recess penetration," *Surface Mount Technology*, February 2002, pp. 15-18.

[111]   Para Tech Coating, "Typical engineering properties of commercial parylenes," http://www.hpetch.se/paratech/parylene_properties.html, as viewed on February 13 2002.

[112]   Parylene Coating Services Inc, "Properties of parylene," http://www.paryleneinc.com/images_s/%20PDS%20Dimer%20International.pdf, as viewed on February 13 2002.

[113]   Donges, Bill, "Conformal Coating Optimization," *Surface Mount Technology*, August 2001.

[114]   Graedel, T.E., Leygraf, C., *Atmospheric Corrosion*, John Wiley & Sons, Inc.: New York, 2000.

[115]   Technical Follow-up Bulletin Number 212600011, Airbus Industries, June 1993.

[116]   Service Bulletin Number M21-3341-037-00, Honeywell, Inc., February 1997.

[117]   Taylor, S.R., "Assessing the moisture barrier properties of polymeric coatings using electrical and electrochemical methods", *Proceeding IEEE Transactions on Electrical Insulation*, vol. 24(5), 1989. pp. 787-806.

[118]   Comizzoli, R.B. , Frankenthal, R.P., Milner, P.C., Sinclair, J.D., "Corrosion of electronic materials and devices", Science, vol. 234, 1986, pp. 340-345.

[119]   Hnatek, E.R., *Integrated Circuit Quality and Reliability*, Marcel Dekker, New York, 1995.

[120] Belton, D.J., Sullivan, E.A., Molter, M.J., "Moisture sorption and its effect upon the microstructure of epoxy molding compounds," *Proceedings IEEE/CHMT International Electronic Manufacturers Technology Symposium*, 1987, pp. 158-169.

[121] Belton, D.J., Sullivan, E.A., Molter, M.J., "Moisture transport in epoxies for microelectronics applications," *Polymers for Electronics Packaging and Interconnection*, American Chemical Society, 1989. pp. 286-320.

[122] Aronhime, M.T., Peng, X., Gillham, J.K., "Effect of time-temperature path of cure on the water absorption of high Tg epoxy resins," *Journal of Applied Polymer Science*, vol. 32, 1986. pp. 3589-3626.

[123] Best, M., Halley, J.W., Johnson, B., Valles, J.L., "Water penetration in glassy polymers: experiment and theory," *Journal of Applied Polymer Science*, vol. 48, 1993. pp. 319-334.

[124] Zheng, Q., Morgan, R.J., "Synergistic thermal-moisture damage mechanisms of epoxies and their carbon fiber composites," *Journal of Composite Materials*, vol. 27(15), 1993. pp. 1465-1478.

[125] Day, D.R., Shepard, D.D., Craven, K.J., "A microdielectric analysis of moisture diffusion in thin epoxy/amine films of varying cure state and mix ratio," *Polymer Engineering and Science*, vol. 32(8), 1992. pp. 524-528.

[126] Maffezzoli, A.M., Peterson, L., Seferis, J.C., "Dielectric characterization of water sorption in epoxy resin matrices", *Polymer Engineering and Science*, vol. 33(2), 1993. pp. 75-82.

[127] Gupta, V.B., Drzal, L.T., Rich, M.J., "The physical basis of moisture transport in a cured epoxy resin system," *Journal of Applied Polymer Science*, vol. 30, 1985. pp. 4467-4493.

[128] Dycus, D.W., "Moisture uptake and release by plastic molding compounds □ its relationship to system life and failure mode," *Proceeding IEEE*, 1980. pp. 293-300.

[129] Gale, R.J., "Epoxy degradation induced Au-Al intermetallic void formation in plastic encapsulated MOS memories," *Proceeding IEEE International Reliability Physics Symposium*, 1984. pp. 37-47.

[130] Gallo, A.A., "Effect of mold compound components on moisture-induced degradation of gold-aluminum bonds in epoxy encapsulated devices," *Proceeding IEEE International Reliability Physics Symposium*, 1990. pp. 244-251.

[131] Khan, M.M., Fatemi, H., "Gold-aluminum bond failure induced by halogenated additives in epoxy molding compounds," *Proceeding IEEE International Symposium on Microelectronics*, 1986. pp. 420-428.

[132] Lum, R.M., Feinstein, L.G., "Investigation of the molecular processes controlling corrosion failure mechanisms in plastic encapsulated

semiconductor devices," *Proceedings of the 30th IEEE Electronics Components Conference*, 1980. pp. 113-120.

[133]   Gallo, A.A., "Effectors of high temperature reliability in epoxy encapsulated devices," *Technical Paper-Delco Electronic Materials Division*, 1993. pp. 1-27.

[134]   Berg, H.M., Paulson, W.M., "Chip corrosion in plastic packages", *Microelectronic Reliability*, vol. 20, 1980. pp. 247-263.

[135]   Dunn, C.F., McPherson, J.W. "Recent observations on VLSI bond pad corrosion kinetics," *Journal of the Electrochemical Society: Solid State Science and Technology*, vol. 135(3), 1988. pp. 661-665.

[136]   Hopfenbergh, H.B., *Permeability of Plastic Films and Coatings to Gases, Vapors and Liquids*, Plenum Press: New York, 1974.

[137]   Crank, J., Park, G.S., *Diffusion in Polymers*, Academic Press: London, 1968.

[138]   Comyn, J., *Polymer Permeability*, Elsevier: London, 1985.

[139]   Boa, T., Tanaka, J., "The diffusion of ions in polyethylene," *Proceedings of the Third International Conference on Properties and Applications of Dielectric Materials*, 1991. pp. 236-239.

[140]   Manzione, L.T., *Plastic Packaging Of Microelectronic Devices*, Van Nostrand Reinhold: New York, 1990.

[141]   Pecht, M.G., Nguyen, L.T., Hakim, E.B., *Plastic-Encapsulated Microelectronics*, John Wiley & Sons: New York, 1995.

[142]   Zhang, H., Davidson, W., "Performance characteristics of diffusion gradients in thin films for the in situ measurement of trace metals in aqueous solution," *Analytical Chemistry*, vol. 67(19), 1995. pp. 3391-3400.

[143]   Zamanzadeh, M., Liu, Y.S., Wynblatt, P., Warren, G.W., "Electrochemical migration of copper in adsorbed moisture layers," *Journal of Science and Engineering Corrosion*, vol. 45, no. 8, August 1989.

[144]   IPC Publication IPC-TR-476A, "Electrochemical Migration: Electrically Induced Failures in Printed Wiring Assemblies," Northbrook, IL, May 1997.

[145]   IPC Publication IPC-SM-840C "Qualification and Performance of Permanent Solder Mask," Northbrook, IL, January 1996.

[146]   Hernefjord, Ingemar, *Cut-out from Report about SBGA Soldering Problems Support from US Companies/ Simple Mechanical Tests*, January 29, 2001.

[147]   Ward, Pamela, "Plasma cleaning techniques and future applications in environmentally conscious applications," Sandia National Laboratories

supported by the U.S. Department of Energy under contract #DE-AC04-94AL85000.

[148]   "How clean is clean? Definition and measurement of cleanliness project final report," NCMS Report 0173RE96, National Center for Manufacturing Sciences: Ann Arbor, MI, June 1998.

[149]   Henry, Mark, "A Methodology for Determining Root Causes of Failures Due to Contamination of Electronic Assemblies," Master of Science Thesis, University of Maryland, 1999.

[150]   Straus, Isidor, (2000), "Understanding the NEBS airborne contaminant requirements," http://www.conformity.com/0008reflections.html, accessed on 1 July 2002.

INDEX